EGI

Animals for Schools

Animals for Schools

The choice, care and educational value of animal life in schools

BY

Terry K. Mills, F.L.S., F.R.Z.S.(Edin.)

and

Jill A. Keeling, B.A., F.L.S.

HARRAP LONDON

First published in Great Britain 1973
by George G. Harrap & Co. Ltd
182–184 High Holborn, London WC1V 7AX

ISBN 0 245 51853 3

Composed in Photon Plantin and printed
by Redwood Press Limited, Trowbridge, Wiltshire
Made in Great Britain

To all 'our' schools throughout the country whose enthusiasm and support have changed into reality our ideals for the Centre for Educational Zoology.

Acknowledgements

The authors would like gratefully to acknowledge the help of the following Authorities, schools and individuals whose assistance has been invaluable in the production of this book:

Bedfordshire Education Authority;
Doncaster Education Authority (Yorkshire: West Riding);
Inner London Education Authority;
Rotherham Education Committee (Yorkshire: West Riding);
Hexthorpe Junior School (Doncaster) Mr McKenzie, staff and pupils;
Paxton J. M. & I. School (South London) Mrs Morley, staff and pupils;
Pulloxhill County Primary School (Bedfordshire) Mr Bransom, staff and pupils;
Saint Albans J. M. & I. School (Wickersley, Rotherham) Mr McGreedy, staff and pupils;
Sir John Lillie Junior School (Fulham) Mrs Bates;
Bob Bray for many of the photographs;
Alan Short (fourteen years old) for the humorous line drawings;
Lyn Stephens for burning the midnight oil *and* the candle at both ends in the pursuit of factual information for the religious education section;
Brian Wright for the many hours he spent in producing the children's artwork for the walls of the animal room;
and last but by no means least, Frank Waters of Harraps, who has not only given us every help and encouragement but also withstood many trials and tribulations on our behalf.

Contents

Introduction

Although we are often assured that the fear of animals so common in adults is 'instinctive' and 'inborn', this does not, in fact, appear to be the case. Pop a friendly snake into the pram with a six-months-old baby and watch reactions—the baby will be delighted to greet and fondle this new friend leaving only the mother to exhibit all the symptoms of screaming hysteria! I am not suggesting that you actually try this experiment—certainly not if you value the friendship of the mother in question—but it must have

Pop a friendly snake into the pram

become apparent even to teachers who are restricted to one age group in one school that young children are in general on far better terms with other animals than are older children and adults. For many years now I have been lecturing in schools to groups ranging from nursery up to sixth form secondary, and it is most noticeable how fear and revulsion towards at least some members of the animal kingdom increase steadily as the inhibitions of ignorant parents condition the children's minds to what is basically an unnatural and undesirable attitude.

In view of this natural affinity between young children and animals, it seems to us that a great deal more use could be made of them as a teaching medium than is generally thought to be the case. While nature study, biology, and the 'facts of life' are quite clearly connected with, or actually devoted to, the study of animals, even the most unpalatable of subjects can be 'sugared' if they are related in teaching to this particular field of study—a child with a mind rigidly closed to arithmetic, for instance, will probably be delighted to count a newly weaned litter of mice (three dark brown, two grey and one fawn, making six in all). In my own family I had a case of a slow reader who refused to show any intelligence at 'the cat sat on the mat' level, but learned to read in double-quick time when presented with a book whose contents he was really interested to learn. In point of fact, every subject up to higher university level can, if it be so wished, be related to animal life—hardly surprising when one remembers that Man is himself a member of the animal kingdom and all his painfully acquired knowledge is as much a part of that kingdom as the most elementary behaviour patterns normally ascribed to instinct or motor-learning. Obviously the range of possibilities in this field is too vast to be covered comprehensively in so slender a volume; whereas the types of animal suitable for the classroom are fairly limited, their educational potential is not, so all we can hope to do is make a few suggestions to encourage those who read them to explore further on their own initiative. Equally obviously, except in the case of very few exceptionally animal-minded pupils, it would not be sensible to attempt to base *all* the teaching in a class on this same theme; the effect would very shortly prove just as boring as the repetition of tables and the cat sat on the mat. If, however, the 'stickier' subjects are first chosen for this type of treatment, excellent results may well be obtained, and a good starting point, well calculated to rouse the interest of even the dullest child, is a statement of the indisputable fact that virtually all the greatest achievements of science, of which mankind is so inordinately proud, were first evolved by lower members of the animal kingdom long before he learned to stand on his hind legs, make fire or wield weapons. This will be dealt with in greater detail later on; first let us see just what types of animal, because of their size, diet, habits, etc., make the most suitable classroom pets.

Suitable animals and their care

It is a very difficult task to define what constitutes a suitable animal or collection of animals to be kept in school without being able to discuss the finer points of animal husbandry and individual educational potential with every teacher wishing to keep examples of living organisms in the classroom. The following is intended merely as a guide to and sample list of the varying types of creatures suitable for use in primary education, and is compiled in a manner similar to that of many of our introductory lectures in schools, so that we can see the development of animal life as if on the balance of an evolutionary scale.

WATER INVERTEBRATES

Studying pond life can be great fun and very absorbing, whether it be out in the field or in an aquarium in the classroom. It is always best to collect your own specimens if possible because you can then see at first hand the sort of natural conditions which must be created if they are to live a full and natural life. With many hundreds of thousands of suitable examples to choose from it is very difficult to select an appropriate cross-section; however, in making this attempt I hope to stimulate an interest in the vast and fascinating world of the small water creatures.

Pond-dipping can be fun

HORSE LEECH (Haemopsis sanguisuga)

Contrary to popular belief, not all leeches are bloodsuckers—so, with about 300 species to choose from, and taking into consideration our purpose in collecting a few specimens, together with their care in the classroom, I think it best not to suggest one of the bloodsucking varieties. The Horse Leech is a very interesting and unusual member of the annelids and requires the minimum of attention.

Leeches have a very wide distribution, being found in most parts of the world except where conditions make life a practical impossibility, such as in desert areas. These strange creatures prefer very slow running rivers and streams in which to live, and occasionally can be found out of water hiding beneath rocks, etc. near the river bank. Because of the manner in which they move around and feed, Horse Leeches find conditions most satisfactory if the water is not too deep and the river bed is well covered with rocks and stones, with vegetation reasonably abundant.

Leeches have a sucker at each end of their bodies, which they use to grip not only food but also various objects in order to assist in their mobility. They can, for example, 'walk' over a rock or 'climb' a water plant simply by attaching both suckers to the object of their choice and then removing the posterior sucker, passing the elongated body through 180°, and again grasping with the posterior sucker before releasing the anterior sucker and repeating the process—rather like doing backward flips in gymnastics. If the leech wishes to travel through the water, it will release both suckers and flatten its body, then, with an undulating motion, propel itself through the water in the direction of its choice. Feeding presents little problem to this annelid, which will eat earthworms, snails, various types of larvae and many forms of insect life, by grasping the prey with the anterior sucker which also contains the mouth parts.

As these creatures are hermaphroditic, one does not require a 'true pair' for breeding purposes. The eggs produced are either buried in the mud of the river bed or attached to water plants or submerged objects. If conditions become very extreme—i.e., a stream dries up—leeches will often burrow into the mud and aestivate (a temporary state similar to hibernation), and if food becomes unobtainable for some reason or other leeches can live without it for a considerable time.

In the classroom leeches should be kept under cool conditions, away from direct sunlight and radiators. The aquarium need not be very large (45cm × 25 cm is more than sufficient), and the bottom should be lined with about 4cm of mud taken from the same area of the stream in which the leeches are to be found. Next add a scattering of coarse gravel, some small rocks and pond weed, and fill it up with stream or rain water (not tap water) before leaving it for a couple of weeks to settle.

If you find your leeches attached to rocks, do not try to pull them off; providing that it is not an enormous boulder, remove the rock complete with leech and place it in the aquarium. Feed on a natural diet of earthworms, snails and insect life.

CADDIS FLY LARVA (*Trichoptera sp.*)

Although the caddis fly larva is familiar to most of us, very few, I feel sure, would recognize a caddis fly if we met one, in fact we tend to overlook the adult form altogether, so fascinating is the infant! In point of fact, the caddis fly is a moth-like creature, living near freshwater wherein it lays its eggs. As is the case with many insects, the adult form is not only inconspicuous but very short-lived, while the larva and pupa can survive for the best part of a year.

The caddis fly's main claim to fame lies in his building propensities, for he constructs a most intricate little fortress to guard his soft body against his enemies. Using silk produced in his own body as the welding agent, he utilizes all manner of scrap material from his immediate surroundings in his architectural exploits, a proceeding not only economical but also well calculated to ensure a perfect match with his environment—the camouflage nets so laboriously constructed by Women's Institutes during the Second World War look pretty silly beside the little caddis's efforts.

These animals can easily be collected by yourself and your pupils, as most streams seem to abound in them during the spring and early summer; all you have to do is paddle into the shallow water, pick up the nearest stone and you will probably find a number of 'twigs' or little sausages of sand adhering to the underside. These are in fact our caddis fly larvae, as their indignantly waving arms and heads will tell you when you take them from their moorings.

Housing these insects presents no problem at all; anything from a jam jar to a proper aquarium tank will suit them nicely, provided that you guard against overcrowding which can reduce the oxygen content of the water to a dangerously low level. Feeding is probably best achieved simply by changing the water from time to time, replacing the old with a fresh supply from the original natural source; this is likely to contain various minute forms of animal life together with scraps of greenstuff, etc. These specimens are some of nature's dustmen and therefore quite happy scavenging among other animals' leavings.

WATER BOATMAN (Notonecta glauca)

This fascinating beetle-like insect is an ever-popular favourite with children as he skulls about ponds with the curiously flattened limbs which so resemble oars that they have earned him his very name. I suppose the most surprising thing about this particular species is that, for reasons best known to themselves, they prefer to swim on their backs, a habit which has brought them the title 'back-swimmers' in America.

These can best be housed in our pond life tank—although some care must be taken in choosing their companions as they can give quite a sharp sting with their little 'beaks', and actually feed upon smaller forms of insect life.

DYTISCUS BEETLE (Dytiscus marginalis)

The Dytiscus or Great Diving Beetle is one of the most savage inhabitants of our ponds and lakes, for he is a voraciously carnivorous species, attacking animals much larger than himself in his lust for flesh. Even quite large fish are not safe from his attentions, and he is the bane of many would-be fish keepers. If then, you decide to keep one in the classroom, he should be given separate accommodation, and it is also important that his quarters be fitted with a roof, as he is an excellent flier and might well join some of your other pets uninvited given half a chance to do so. He also has the distinction of being the largest aquatic beetle in this country, so will easily be recognized if you meet him on one of your nature walks.

Although the sexes are fairly easily distinguishable by the fluting on the elytra, I have never met anybody who has succeeded in breeding these animals in confinement—still, there is a first time for everything, so why not try to find a true pair and try your luck? Like many diving beetles, the Dytiscus raises his tail to the surface for air, and the air bubble he takes back under water with him can often be clearly seen—yet another case of other animals forestalling man, for the beetles thought of carrying their air supply below with them long before the first diver ventured down with his canister of oxygen.

MOLLUSCS

Water snails and fresh water mussels are the most likely molluscs for you to find on your pond-dipping explorations, and they are welcome additions to any cold water tank because of their feeding habits. Several species of snail are to be found in our ponds and streams, the most common types being the Planorbis and Lymnaea, and all are gastropods—that is, single-shelled molluscs which literally 'march on their stomachs', for the mouth is actually situated in the foot. With only one known exception which reproduces parthenogenetically—i.e., from a single specimen, no mating being required—snails are hermaphrodite, both sexes being represented in each individual, so if you collect any two snails of the same species you are likely to increase your stock quite quickly. Snails are, of course, vegetarians and, although they may reduce the abundance of any aquatic plants you have rooted in your tanks, they will also do you a great service by industriously munching the inevitable growth of algae off the glass—thus relieving you to a large extent of the rather boring task of scraping it off yourself.

Mussels, on the other hand, are filter feeders, and their task in your aquarium is to filter and purify the water for you. Obviously only comparatively little in the way of food particles can be found in suspension in the water, so a single mussel only will be able to find a living in a 62cm tank. Swan Mussels, the type you are most likely to find, are among the slowest moving of all animals, and it may take one a week or more to plough its way from one end of the tank to the other—you can see the marks of its progress as a definite furrow across the floor covering of sand or gravel. Biologically these are interesting animals as they provide an example of mutually beneficial behaviour between two very different species; the young larva (glochidium) of fresh water mussels undergoes part of its development as a parasite on fish, while the bitterling in her turn likes to deposit her eggs within the open shell of pond mussels, an excellent instance of that ecological inter-reliance between widely dissimilar creatures which is surprisingly common in nature.

In any aquarium one of the major dangers is that some smallish animal will die unobserved and the decomposing body pollute the water, thus more than likely causing the death of all the other inmates of the tank. In a mixed pond-life tank this risk is somewhat reduced as our caddis fly larvae, leech, or even water boatmen may well clear away the corpse before it has done any damage, while the mussel will be working overtime filtering away the remains. Should the mussel itself die, however, you may be in real trouble, for this cadaver will be too large to be disposed of quickly and hygienically. Furthermore, the mussel is so ultra-lethargic that its demise may well go unnoticed for several days, by which time the damage will have been done. If, therefore, you see no signs at all of life, such as partially opened shell which slowly closes upon disturbance, or the remains of a plough-track in the sand, it is advisable to lift the mussel out of the water and take a quick sniff at it—the only certain way that I know of ascertaining whether it is indeed living or dead!

LAND INVERTEBRATES

Just as the word 'animal' to most people means 'mammal', so, I suspect, the word 'invertebrate' generally suggests to them an insect; so perhaps we had better start with a few species of this particular type of animal especially suitable for school. First of all, what exactly is an insect? I suppose its most important and invariable feature is its six legs, which immediately distinguish it from the arachnids, or spider family, otherwise very similar at first sight. Most insects have bodies divided into three distinct sections —*head; thorax* or chest, which carries the legs and/or wings; and *abdomen* containing the digestive, excretory, and reproductive organs. The wide variety to be found among the insects is unrivalled in any other branch of the animal kingdom, for they are found in every available habitat both on land and in water, and in most cases each individual takes on at least three distinct forms during its lifetime, emerging from the egg as a caterpillar or other type of grub, then pupating into a chrysalis, and finally appearing in the adult form as a butterfly, bee, bluebottle or what have you. Some insects are even known to have as many as twenty-one different forms in their life cycle—e.g., *Phylloxera quercus*, the oak phylloxera. It is known that insects existed in vast numbers three thousand million years before man first appeared on this planet, and scientists estimate that there are now between two and four million species although only some 625,000 have so far been described and classified! So at least you have plenty of choice if you decide to keep insects, although those available, suitable and large enough to be seen without the aid of a microscope are rather more limited in their numbers!

INDIAN STICK-INSECT (Carausius morosus)

I suppose that there is only one insect which is kept as a pet in appreciable numbers, and this is the common Indian Stick-Insect. These are often to be seen for sale very cheaply (about 5p each) in pet shops, are easy to keep, and often all *too* easy to breed. Also they are relatively uninteresting, as often tends to be the case with the more common and easily obtained and maintained animals.

Stick-insects are so-called because of their incredible camouflage, both form and colouration giving them an almost uncanny resemblance to the twigs among which they live, a resemblance enhanced by their aptitude for 'feigning death' and remaining absolutely motionless with their legs folded alongside their bodies for long periods. Even if you know they are there you can gaze into a well furnished tank of these insects for several minutes before you succeed in spotting a single one! The Indian Stick-Insect is leaf green or brown in colour, usually with small red markings at the base of the foremost pair of legs. It has two long and slender antennae, or feelers, which help it to find its way about, as its compound eyes do not appear to be very effective. This is one of the insects which do not undergo 'metamorphosis', a series of changes through completely different forms; instead it is hatched from the egg as a minute replica of the adult—a hair-like half-inch hatchling may develop through several moults of its exoskeleton into a thickset adult anything up to 10 cm long. Stick-insects are strictly vegetarian in diet, and this particular species is especially easy to keep because the food plants, privet and ivy, are readily available throughout the year. It is also easy to breed because it is 'parthenogenetic'—in other words, the young develop from unfertilized eggs, so that you need buy only one female to build up an ever increasing stock of her progeny. I specify a female because, strangely enough, although no mating is necessary for the reproduction of this species, the odd male does crop up from time to time—presumably nature's way of strengthening the stock by an occasional dose of cross-breeding. The male is easily distinguished, however, being of a much more slender build than his myriad wives.

Should you decide to keep stick-insects in school, the equipment you will need can be just as homely or as sophisticated as you wish, in fact anything from a two pound jam jar to the most up-to-date vivarium which will cost you the earth from your local pet shop will serve your purpose in housing them. Whichever 'standard of living' you decide upon, I would strongly advise you to choose a 'top-loader', as any container with an opening at the side tends to land you in trouble once your specimens have got under way as a breeding colony, for they will all spill out in every direction whenever you renew their food supply. At the same time, too narrow a top opening is also to be avoided because, to ensure freshness, the twigs of privet or ivy should stand in a jar of water and be renewed as soon as they either wither or are eaten bare of leaves, and you can imagine that trying to extract or reintroduce a jar of water through an opening that will barely take your hand can be difficult in the extreme. I have found large sweet jars, easily obtainable for a few pence, quite suitable as what you might call 'cottage-type' dwellings, the tops being covered with muslin or some similar material, held in place by an elastic band, to ensure adequate ventilation. One hazard which you must always be on the look-out for as your stocks increase and multiply is overcrowding, which often induces the insects to bite off each other's limbs. Certainly these can usually be regenerated (grown again), but the effects are far from pleasing either to you or to the animals themselves, so do take care to thin them out from time to time, either by selling them back to your local pet shop, giving them to your friends or pupils, or using them as an excellent food supply for such other animals as lizards which you may also keep.

The eggs of the Indian Stick-Insect look very much like radish seeds (yes, even at this early stage an effective vegetable camouflage is adopted!), being perfectly globular, often with a small spot of a different colour somewhere on the surface—in fact, it is only by their shape that you can easily distinguish them from the droppings, which are of a similar dark-brown colour but much more elongated. If you wish to breed stick-insects some so-called experts will tell you that you should laboriously collect these eggs from among the conglomeration of faeces, fallen leaves, and general detritus in the bottom of the tank, place them in a match box or other small container, and keep them in a fairly warm, damp place until they hatch. The tiny, fragile babies which appear in due course should be gingerly removed by means of a camel-hair paint brush to yet another container, where they must be fed upon the finely shredded leaves of the appropriate food plant. All of which is a great deal of unnecessary work, involving a number of very real hazards. As the eggs may take anything from about six weeks to six months to hatch, it is more than likely that you will forget all about them and the hatchlings will be born only to die of starvation, while the removal of the infants, however carefully performed, is liable to cause injury in some degree or other to quite a large percentage of them. Personally, I prefer a much simpler and more natural method, which I achieve by simply leaving everything which falls to the bottom of the tank strictly alone. In a surprisingly short time you will amass quite an appreciable depth of 'loam' which, of course, provides an ideal incubator and nursery.

As for 'finely shredded leaves'—no kindly soul goes trotting round the forests and jungles of India painstakingly shredding food for all the millions of baby stick-insects which are born there annually, and a fair percentage of them seem to do all right! Just as a point of interest, I am quite gratified to note that while I must have bred thousands if not millions of the more common species of stick-insects by this method, the wielders of camel-hair paint brushes apparently count their successes only in hundreds.

As I remarked earlier, the common Indian Stick-Insect is one of the least interesting of the phasmids, as this group of animals are called. There are many other types which are nearly as easy to keep yet far more attractive, although unfortunately they are also more difficult to obtain. We do have surplus stock for disposal from time to time, however, if you are really interested in obtaining any.

CORSICAN STICK-INSECT (Clonopsis gallica)

Similar to the common Indian Stick-Insect, but with short antennae, this species appears in green, brown, and pink shades. It is very hardy, and has actually been introduced successfully into certain areas in the south of this country. Food plants: bramble, raspberry, rose.

WINGED STICK-INSECT (Necroscia sipylus)

Hailing from Madagascar, this beautiful insect is pale straw colour when adult, though bright green in infancy. The large lacey wings do not appear until several moults have taken place, but when they do the creature is capable of short flights which may be sustained for nearly the length of the classroom, so be sure to shut the windows before you open the tank containing this particular species. Food plants: bramble, raspberry, rose.

THORNED STICK-INSECT (Clitumnus extradentata)

A native of Australia, this species is especially twig-like in appearance, being slender and of a woody brown colour with small thorny excrescences along the body and limbs. Reproduction is sexual, the females being easily disinguished by their much fatter bodies. Food plants; bramble, rose, raspberry.

JAVAN STICK-INSECT (Orxines macklotti)

This attractive creature is much more conventionally insect-shaped than the species previously mentioned, the females especially (here again two sexes are represented) having the swollen, tapering abdomen we associate with animals of this type. In colouration they are speckled olive green and brown, and when they fly, which is seldom, the wings are revealed in a splash of deep indian red. Food plant: rhododendron.

AUSTRALIAN LEAF-INSECT (Extatosoma tiaratum)

I cannot help thinking that in the wild members of this species must favour a diet of some type of cactus, because they remind me more than anything of animated opuntias. They can grow to an amazing size (I have heard of specimens weighing up to a quarter of a kilogramme) and have curiously flattened and distorted bodies which give them an appearance quite out of this world. Unlike the other species mentioned they normally carry their tails curled over their backs, and like to stand swaying from side to side, rather like grotesque elephants. They can grasp remarkably tenaciously with their claw-like feet, and are not above slashing out at your hand when you pick them up—not that they can make much impression on human skin. In colouration they are black with reddish heads as hatchlings, but later become green or brown, and only the males have rudimentary wings on which they can flutter for short distances. Food plants: bramble, rose, raspberry.

COCKROACH

Most people, I know, will shudder at the thought of these insects being kept in school, but surprisingly enough some of them are not only very interesting to study but also very beautiful! Have you ever really looked at an ordinary cockroach—the species which you might be horrified to find in your own kitchen? They really are lovely creatures, if you can bring yourself to regard them dispassionately, with shiny brown, slenderly built bodies and delicate 'antennae' or feelers always vibrating in a ceaseless endeavour to discover all they can about their surroundings. Their flattened bodies are wonderfully well adapted to enable them to scuttle out of sight of their enemies below stones and fallen branches, or underneath the edge of the linoleum when they have chosen to share some human habitation! Like most cold-blooded animals, cockroaches like warmth, so it is hardly surprising that those living in our rather inclement climate tend to seek out the artificial sources of heat to be found in houses and factories. Being nocturnal (coming out only at night), they are extremely elusive insects, and many a household may have unwittingly harboured them for years before some member of the family seeking a midnight cup of tea surprises them at their forays long after everybody else is in bed.

Cockroaches are the most primitive of all winged insects, and also among the oldest insects found as fossils; both of which facts should surely earn them a place in our interest if not in our affection! They are also valuable in the general economy of nature in that they play an important role as dustmen, for they are omnivorous in diet, that is to say they will eat virtually anything at all, which of course includes a good deal of waste matter which would soon build up into a vast accumulation of litter without their tireless efforts to dispose of it. Certainly we humans may not be best pleased if our neighbours the cockroaches choose to regard our books and papers, shoes and clothes as so much litter or food, but the fact remains that they do excellent service as vacuum cleaners, zealously clearing up all the crumbs of food and bodies of small dead creatures to be found on their nightly prowlings.

If you are lucky enough—yes, I do mean that, lucky enough—to have a friend whose kitchen is 'plagued' with cockroaches, you can catch some fairly easily if you decide to start a colony. Simply bait a large, deep jar with a little bread and jam, and await results. After a couple of nights you are almost certain to find quite a number of the insects safely captured in your jar, having gone into it after the bread and jam, and been quite unable to climb out again. Do not leave your trap too long, though, or you will find only one fairly fat cockroach and a number of discarded limbs, all that remains of his less robust cousins, for in times of famine these creatures are not above indulging in a little cannibalism in order to assuage their own hunger! Once caught, cockroaches are easily housed in an ordinary fish tank, for they seldom fly and will find the overhanging 'lip' at the top of the tank quite an insuperable obstacle if they attempt to escape. To be absolutely certain that you have no escapees, however, fit on a top of perforated zinc, or even a sheet of glass, for the irregularities of the metal rim at the top of the tank will allow the passage of quite sufficient air. Put about 3 cm of bran in the bottom of the tank and add a few pieces of bark or flat stones for your specimens to hide under during the hours of daylight, and after that you can feed them on nearly anything you like—bread, vegetables, bits of meat, cheese rind—the choice is yours! Water is best supplied by means of a scrap of cotton wool, rung out in clean water and suspended just above the level of the bran, so that it remains uncontaminated and does not damp the whole floor covering.

Cockroaches prefer to be warm, as already mentioned, so near the class radiator would be a good place to keep your tank. Usually these insects settle down well in confinement and soon start to breed, the young being very different both in shape and colouring from the adults. Most species lay eggs, but some are 'ovoviviparous', that is to say, the eggs are kept in the mother's body and actually hatch there, the young emerging alive from her cloaca or anal vent. Cockroaches are used a great deal in schools and colleges for dissection, as they are sufficiently large and their organs clearly enough separated to give the student a fair idea of the workings of an insect's body, so when any of your specimens die you may be interested to open them up and see 'how they tick', or pass them on to the biology master of your local secondary school. Be warned, though; cockroaches carry a thick layer of yellow fat within the abdomen, which has a most unpleasant smell, so you may well have to hold your nose during your quest for knowledge!

The Common Cockroach (*Blatta orientalis*) is the species you are most likely to find yourself in this country—only two species actually live here anyway—but, as is the case with the Indian Stick Insect, the most easily obtainable is by no means the most attractive. The Giant Death's Head Cockroach (*Blaberus cranifer*) from America is far more interesting. In this species the very young are tiny creamy coloured creatures, roughly oval in shape; the 'adolescents' are almost circular when viewed from above, and heavily spotted with dark brown on a golden-cream background; and the adults have the typical cockroach shape and the handsome markings which give them their name. These seem to become rather more diurnal than most other species and thus are more easily observed if you decide to keep a few. Also, of course, they are a great deal larger than the Common Cockroach, some growing up to at least 5cm in length, so that as exhibits they are much more eye-catching and spectacular.

PRAYING MANTIS (Sphodromantis gastrata)

Another member of the cockroach family 'Dictyoptera' is that rather mysterious creature the praying mantis. Mantids are not the easiest of insects to keep in captivity because they are very delicate little animals and many of them have a very short lifespan.

There are about 800 species of mantid known to science, all of which live in the warmer countries of the world. The type that I have had most success in keeping is the species *Sphodromantis gastrata*, which comes from South Africa and grows to a length of about 7½cm. The praying mantis gets the 'pr*a*ying' part of its name from the manner in which it holds its first pair of limbs, and not from the fact that it pr*e*ys on other animals (mostly other insects). Another very interesting thing about the praying mantis is the number of parts into which its body is divided. We usually think of insects as having only three distinct parts to the body, namely the head, thorax and abdomen, but the mantid's body is divided into four parts. The extra section is called a pro-thorax, which we would probably know better as a neck. This neck allows the creature to move its head up and down and from side to side just as our neck does. There are two feelers or pre-oral antennae on its head and two pairs of gossamer thin wings are concealed along the dorsal surface of its abdomen.

I think that a praying mantis is one of the most beautiful of all the insects that I have ever seen, and to see one of these magnificent creatures at rest on the branch of a tree, with its head turned heavenwards and the first pair of legs held in a worshipful attitude of prayer, can only be described as sheer delight.

The mantis uses its two long whisker-thin feelers to help it to find its food; as these feelers are very sensitive to vibrations in the air they can detect the smallest of animals moving around nearby. Once the mantis has detected its prey with these feelers, the large compound eyes are used to see exactly where it is. Next it moves slowly towards the unsuspecting creature by using its four hind limbs to propel itself while keeping the much larger pair of front legs held erect near its head. When it reaches its potential meal, the mantis shoots out these forelimbs and seizes the insect firmly in its grasp. Once caught, the creature finds it almost impossible to escape, for if we examine the mantis's forelimbs closely we can see rows of sharp spines which close up against and between each other, and at the very front of each large grasping limb, we can see a very long curved spine. It is with these spines that the mantis makes its first strike, using them like daggers to spear the food. The mantis is not equipped with any poison or paralyzing sting as a wasp is; once caught, the prey is eaten alive. Small birds are also on the menu of the praying mantis, and while I would agree that this does not sound very nice, it does illustrate the speed with which a mantis can grab the things it wants. If attacked by a lizard, large bird, or other enemy, the praying mantis will make an often futile bid for survival by trying to fend off the assailant by using the two front legs in the manner of swords.

Like the female spider, the female praying mantis usually kills her mate after mating. This may sound rather barbaric and ungrateful, but it is often very necessary because, in some cases, unless the male is killed and eaten, a hormone necessary to aid the fertilization of the female's eggs cannot be released, and without this hormone her eggs will be infertile. Before the female actually produces the eggs, she spins a silky cocoon into which about ten eggs will be laid. Once the eggs are safely inside the cocoon the female will seal up the opening with more of her silk, and tie the cocoon to the branch of a tree.

Providing a praying mantis with a home is quite an easy task, as the only materials required are a round cake tin, a sheet of very thin clear plastic, some perforated zinc, a little solder and a tube of glue. Remove the lid of the tin and cut a hole in the centre about 14cm–16cm across, removing any sharp edges; cut the perforated zinc a little larger than the hole to give a nice margin for soldering and, working from the inside of the lid, clean the area around the hole before soldering the zinc into place. This completes the ventilation requirements and the tin can now be painted if you wish. Take the plastic, which should be about 45cm long, the width depending on the diameter of your tin, wrap it round the outside lip of the bottom half of the tin and cut to size, leaving about $1\frac{1}{2}$cm spare to provide a seam. Now glue the plastic into the shape of a tube and leave to dry, placing something heavy along the length of the seam to hold the glued surfaces together. When dry, put the plastic tube on to the bottom half of the tin, place the lid on top, and your 'larva case' is complete. It can be heated from below and lit from above, as both heat and light are very important, for the praying mantis is a very delicate animal, and must be kept at a temperature of about 21–24°C.

The mantis does not require very much room, as it seldom takes to the wing—in fact, you will most probably find that it spends most of its time sitting quite still on its branch. The furnishing of the larva case should consist of a gravel floor and a small branch. Feed on flies, moths, caterpillars, etc., when available; you will find that mealworms are a useful standby.

GIANT RED-LEGGED MILLIPEDE (Archispirostreptus gigas)

If you were to go into a garden and turn over various rocks and stones, one of the many creatures that you would find would almost certainly be a millipede. These strange invertebrates are found in a great many countries and come in a variety of sizes. One of the largest of them all is the Giant Red-legged Millipede which is found very widely distributed in South Africa. When

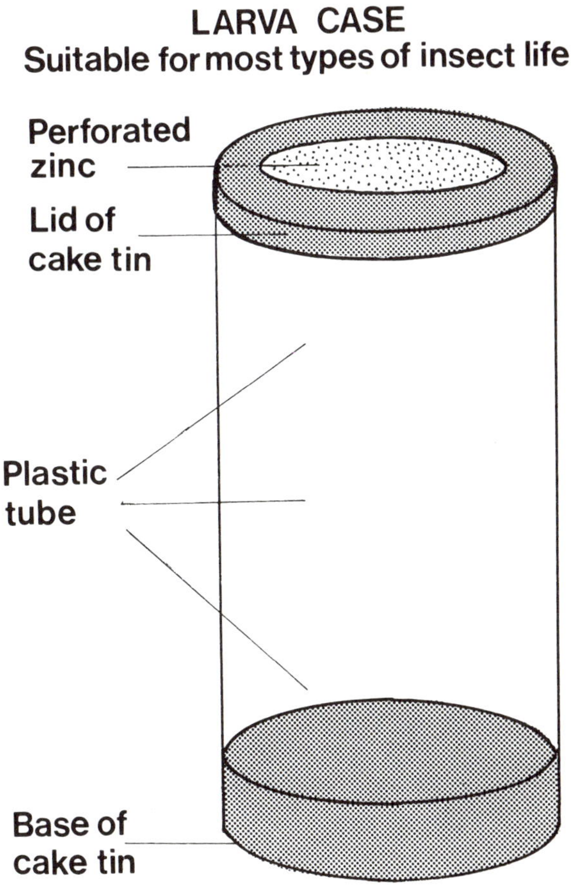

A Larva Case
(suitable for most types of insect life)

fully grown this creature can be as long as 20cm. The exoskeleton, or hard outer skin, is divided into about eighty segments, each bearing two pairs of legs—about 320 legs altogether. 'Millipede' really means 'a thousand feet', but this, of course, is a gross exaggeration intended only to imply that there are a great many legs. Each foot is equipped with very small hooks to enable the animal to grip the surface upon which it is moving—it can even climb up vertical walls or hang on upside down. The senses are generally poor, ears and olfactory organs being entirely absent and the 'eyes' no more than primitive eye-spots barely able to distinguish between lightness and dark. The antennae, however, are quite sensitive, and it is with these that the creature finds its way around in the world.

Unlike their close relatives, the centipedes, which have a poisonous bite, the millipedes are harmless herbivores, feeding at night to escape the attentions of birds, monkeys, lizards, etc., which may eat them. Like the English varieties of millipede, the Red-legged Millipede is to be found hiding under stones, dead leaves, etc., during the day, where he can sleep away from the sun and potential enemies.

If you are thinking of keeping a Red-legged Millipede in school a factor which must be emphasized is warmth, for without a reasonable degree of heat this creature will surely die. 21–24°C during the daytime is sufficient, and this should be reduced to 18–21°C at night. I think it best to house the millipede in a small reptile vivarium or an aquarium tank, but if you use the latter you will need something a little more excape-proof than a normal aquarium hood as these creatures can climb surfaces which are almost completely smooth and are very good escapists. A nice deep carpet of peat is ideal for this animal, but be sure to provide a piece of bark or some similar object for him to hide under. Sexing millipedes is very difficult, but the female is generally bigger than the male and some authorities claim that the absence of legs on the seventh segment is a sign of masculinity.

FISH

One of the drawbacks in writing this book has been the inescapable fact that there is far too much to say. Never yet have I finished a single section without feeling that it was in fact incomplete, and I really ought to expand it for many more pages. It is, therefore, quite deliberately that we have kept the sections relating to fish and birds as short as possible, on the grounds that so much specialist literature has been produced about them that it would be a waste of our all too restricted and therefore valuable space to expatiate upon them to the detriment of other animals or aspects of animal life less thoroughly covered already. Do not be surprised, then, to find no more than some half dozen species of fish dealt with here, nor labour under the delusion that these are the only types we consider suitable to be kept in schools. The possible inhabitants for your classroom aquarium are innumerable, particularly when it comes to stocking a tank of tropicals. I have merely selected a few which seem to me to be of especial value from an educational point of view; further combinations and permutations are up to you.

GOLDFISH (Carassius auratus)

I suppose most aquarists start their career with their first goldfish, probably won by accident at the shooting gallery of a fair, and certainly it is the most common species to be kept as a pet both in bowls and tanks indoors and to decorate outdoor lily-ponds. A member of the carp family, the goldfish is most closely related to the Crucian Carp, and in the wild state is a brownish green colour, which hue is still retained by the fry (young fish) bred in captivity. For centuries the goldfish has been selectively bred by the Chinese and later the Japanese and many weird shapes and colours have thus been produced; every shade from white to deep red through yellow, or a combination of two or more of these, with or without black markings are to be found in goldfish, and some of the strange, almost monstrous forms have to be seen to be believed. Some have long, flowing triple-finned tails (Veiltails), some are jet black in colour and have strange, protruding goggle-eyes (Telescopic-eyed), some are bred entirely without scales (Shubunkins); there are short ones and long ones, thin ones and fat ones—in fact, practically every shape a fish could be. One word of warning; many of the more grotesque types are excessively prone to swimming-bladder trouble, an ailment which causes the sufferer to lose its equilibrium and become quite incapable of swimming in an upright position. This condition can sometimes be cured by reducing the water level until it is barely deep enough to keep the affected fish afloat, but it almost invariably recurs and sooner or later is the inevitable cause of death. So, if you hope to keep your fish to a ripe old age it is advisable to choose the conventional rather than the exotic types.

Accommodation for goldfish presents no problem, as they are quite hardy and can live equally happily in an outdoor pool or in an aquarium tank in the classroom. Do *not*, however, on any account condemn your fish to life in a bowl. Bowls are quite unsuitable on several counts, the most important one probably being that the surface area of the water inside the bowl is quite inadequate for its cubic capacity which means that it is always insufficiently oxygenated. And anyway, how would you like to live in a circular cell, going round and round an endless and unchanging wall for the rest of your life? The floor of the tank can be lined with gravel planted with suitable aquatic vegetation to provide natural aeration, the accumulation of mulm being removed with a vacuum as and when necessary. Always get your tank well settled and established before introducing the fish and, should a change of water become necessary, take great care to ensure that there is no sudden change of temperature as this is very detrimental to them. In fact, it is best to avoid water changes, if at all possible, as the catching up of the fish which these necessitate does them no good at all and can easily damage the mucous covering on their bodies thus giving access to various diseases such as white-spot or fungus. Luckily there are various preparations on the market which can deal with these ailments, in addition to older (and probably better tried) remedies such as a few drops of mercuro-chrome for the former and a pinch or so of common salt for the latter; however, it is obviously better not to get the troubles in the first place, prevention being better than cure, so it is worth while taking a good deal of trouble to provide and maintain a well-balanced tank in which the water will remain sweet and clean almost indefinitely.

Goldfish can be fed upon any of the multitude of proprietory foods on the market, but will probably do best if you give them a preponderance of live food in the form of earthworms, maggots, etc., with the addition of a few bread or biscuit crumbs. Fresh, natural and, above all, living food is advantageous for nearly all fish, and nothing will ever convince me that dehydrated ants' eggs, for instance, have the same food value as their newly gathered counterparts.

If you keep goldfish in an outdoor pool you will find that they tend to go off their food in winter, so it is advisable to feed them really well in summer so that they can build up reserves to last them through the cold months. Do not over-feed though, as this will soon turn the water sour and quite deadly to its inhabitants (this is an even bigger hazard in a small tank). Also the surface of the water must be kept free from ice or it may become deoxygenated during a prolonged frosty spell; this is best done with a small aerator, as breaking the ice already formed causes shock waves which may harm or even kill the fish. Although not strictly a goldfish, the Golden Orfe (*Idus idus*) makes a most decorative addition to an outdoor pool and can grow to a most impressive size. Predators are probably one of the most serious hazards involved in keeping fish out-of-doors, and these range from herons, who can decimate your stock overnight, to Dytiscus Beetles who will prey upon even quite large fish. Frogs are also to be discouraged at the breeding season, as amorous males who can find no mate of their own species are not above clinging to unwilling goldfish and even suffocating them! Frog spawn, however, can safely be left if it does appear, as the tadpoles make a welcome addition to the fish's diet.

MINNOW (Phoxinus phoxinus)

Minnows are included here because they are one of the few British wild animals which can safely be caught in small numbers and transferred to the classroom. The smallest members of the carp family, they inhabit rivers and streams all over the country, and can often be caught in wide-necked bottles left baited with scraps of food in the water; when you retrieve these, lifting them quickly out, you are very likely to find a few fish guzzling away inside.

Unless they can be kept in a really large pond where aeration presents no problem, minnows require the benefit of an aerating pump constantly blowing air through their tank—but they can sometimes be conditioned to accept still water if the degree of aeration is gradually reduced and the tank is not over-populated. Aerators can be obtained for about £1.50, and have the advantage of keeping the water clean as well as oxygenated if they are attached to one of the many filters obtainable from pet shops. I have found the best of these to be in the form of a small plastic box-like container, filled with glass wool and placed in one corner of the tank. These have the advantage of being easily serviced, which is a major consideration if you are hoping to let the children attend to the tank themselves. The filter has to be weighted by a layer of gravel in the bottom, and a small quantity of carbon beneath the filter medium helps to keep the water sweet and pure.

Minnows thrive upon the same diet as goldfish, live food being absolutely essential as they are not a domesticated species.

STICKLEBACKS (Gasterosteus aculeatus and *Pungitius pungitius)*

The Three-spined and Nine-spined Sticklebacks which may well be discovered upon pond-dipping expeditions are excellent subjects for a classroom aquarium, being not only most attractive little fish but also very interesting in their habits. Kept in a well planted tank and provided with plenty of live food such as whiteworm or tubifex, sticklebacks will often breed and thus provide material for our 'facts of life' teaching. It is inadvisable to keep more than one male in a tank, as these are exceedingly pugnacious little fish and will defend their chosen territory fiercely against any intruders; more than one female is quite safe, however, for the males are polygamous and welcome the presence of a harem of wives. In the spring the males acquire vivid breeding colours of red and blue, and construct complicated nests from scraps of herbage, etc., from the surrounding territory; here they invite any females who are willing to present them with eggs which they fertilize immediately they are laid. Only when the nest is full is the urge to reproduce satisfied.

The incubation period for the eggs is between ten and eighteen days, during which time the indefatigable male defends the nest unceasingly, rearranges the 'bedding' from time to time, and ensures proper aeration for his brood by fanning the water with his fins to produce a current over and around them. Even after the youngsters are hatched, father keeps up his duties for several days, often having to protect them from the attentions of their own mothers who are not above sampling them as an item on the menu! As a single female can deposit anything from fifty to a hundred rather large eggs your stock of sticklebacks is liable to increase with alarming rapidity, thus affording you an excellent opportunity to work in the interests of conservation by returning the fry to their natural habitat. For those you decide to rear in the aquarium a diet of liquifry for egg-layers is most likely to be successful, as finding sufficiently microscopic natural food is wellnigh impossible.

COMMON EEL (Anguilla anguilla)

I cannot remember ever having seen a fish of this species offered for sale in a pet shop, but you can often obtain them by visiting a fish market as they are exceedingly tenacious of life and will live for long periods out of water. In fact I know of one which apparently fell off a lorry (literally!) in a London street where it was found and rescued by a schoolmaster who now proudly displays it as one of his prize exhibits! I have myself collected specimens both direct from fishing boats in Essex and from a fishmonger's slab in Amsterdam, and some of them have lived with me for many years in spite of their initial rough treatment. Both zoologically and geographically these are exceptionally interesting fish, and worthy subjects for any classroom aquarium.

Zoologically speaking, these are cartilaginous fish, unlike those previously mentioned which are all bony, and therefore probably older types than the goldfish, minnow, etc. This term means that the skeleton is quite soft, almost like toughened gristle in fact, and probably constituted nature's first move away from the invertebrates towards the animals with backbones. The whole shape of the eel is much more simple than that of other fishes, with only rudimentary fins and the dorsal and caudal fins running together right round the latter end of the body, a sure sign of an early form. The eel has no scales, but only an extra-slippery mucous covering to its skin, which makes it very difficult to get hold of by hand, while the long, sinuous, streamlined body helps it to slither deftly out of your net too—so these are by no means the easiest of fish to handle. They can not only live out of water for quite long periods, but will even travel a considerable distance over-land to get from one stream to another or to return to their native element if some predator catches them and accidentally releases them before they are eaten. One of their most amazing characteristics, shared by the salmon, is their ability to live in both salt and fresh water, an almost incredible state of affairs when one remembers that the law of osmosis requires that solutions on opposite sides of a membrane shall be of similar density; just how an animal whose body fluids are of the correct concentration to enable it to live in fresh water can swim out into the ocean without a radical change in its whole metabolism is little short of a miracle. But this is just what it does, undergoing the change from marine conditions to fresh water and back again not once only but twice during its lifetime. The transparent larvae of the Common Eel are hatched far away in the Sargasso Sea, where they live near the surface of the ocean and spread in all directions. They grow from about 10mm to 75mm long in something like two years, by which time they have found their way into the estuaries of the rivers they are about to ascend. In their thousands they swarm up the rivers and into the streams and rills, sometimes crossing dry land to reach isolated ponds and lakes where they grow to maturity. They reach sexual maturity at anything between 4½ and 8½ years old, the males being rather smaller than the females with a maximum length of some 50cm to the females' 65cm. At this stage yet more changes occur, for the fish put on a great deal of fat, acquire larger eyes, more pointed snouts and pectoral fins, and turn silvery in colour instead of the greenish grey of their infancy, before ceasing to feed altogether and setting out on their marathon return journey to the seas of their birth, where they just have the strength to spawn before their death. Quite a saga—the saga of a diminutive body whose vast journeyings make mincemeat of our steamship timetables!

If you decide to keep these living geography lessons in your classroom, you will need no very complicated equipment. They will do very well either with or without aeration, require no heat, and thrive on a diet composed of nothing more exotic than earthworms. The tank should have sufficient gravel on the floor for them to burrow into, an occupation they favour greatly, prefering to hide during much of the day and swim about at night, and they should not on any account be required to share their quarters as they are extremely voracious and liable to devour any tank-mates you may think of putting in with them. I once made the mistake of thinking I could safely keep an eel with a crayfish, knowing the former to be too slippery to be endangered by the latter; I had a rude awakening when the unfortunate crayfish sloughed, and was promptly gobbled up by the waiting eel.

GUPPY AND SWORDTAIL (Lebistes reticulatus and *Xiphophorus helleri)*

I am lumping these two species together as being probably the easiest of all tropical fish to keep, because they can live together amicably in the same tank, and because they are both live-bearers requiring identical treatment for breeding purposes. If you wish to keep tropical fish you will, of course, require rather more complicated equipment than is necessary for cold-water fish; the tank should be prepared in much the same way as described for goldfish, with a floor of gravel and various plants rooted in it, but in this case a heater, thermostat and thermometer must also be provided to ensure that your specimens are kept constantly at the optimum temperature—about 26°C. These fish will do well on a basic diet of dry fish food (I have found Tetramin products the best), but live food is once more beneficial—expecially if you hope to breed with them. Only the smallest earthworms can be tackled by these small creatures, and whiteworm and tubifex are probably better. Should you feed the latter, however, do make sure that you mince or crush it up well immediately before feeding, as I have known live tubifex to live on inside the gut of the fish that swallowed it and in turn start eating its own devourer!

Guppies may now be considered a domesticated species, for they have been bred in many different types, and even the swordtail appears in both the natural green form and the red artificially produced one. In both cases the sexes are easily distinguished, the male guppies being at once much smaller and more highly coloured than the females, while in the case of the swordtails only the males sport the elongated tail from which their name is derived. You have only to ensure that both sexes are represented in your tank to be sure that breeding will be both frequent and profuse. It is easy to see when the females are gravid as a dark blob becomes clearly visible in the lower abdomen. This is the time when the lady should be isolated if you hope to rear the young, as most fish have cannibalistic tendencies and the babies are likely to be devoured immediately if born in the communal tank. A large jam jar floated in the tank makes an excellent maternity hospital, and the female can be moved back to her old home as soon as you see that she has given birth, for even she is not above fancying 'baby-on-toast' for breakfast. The young are easy to rear on Liquifry for live-bearers, and will grow very rapidly.

SIAMESE FIGHTING FISH (Betta splendens)

In the same tank as your guppies and swordtails can be housed the very spectacular Siamese Fighter, a really beautiful little fish. He does not get his name for nothing, for he is an exceptionally pugnacious character and will not tolerate another male of his own species anywhere near what he considers to be his territory. Fortunately this king-of-the-castle attitude does not extend to other species of fish, and solitary male fighters are excellent denizons for community tanks. Like the sticklebacks, the male fighter builds a nest for his mate's eggs, but in this case he constructs it from air-bubbles he blows himself and floats on the surface of the water. If you intend to breed fighters it is probably best to segregate them from the community tank, as the bubble nests are very fragile and easily broken up if the proud father is constantly dashing about driving inquisitive neighbours away. Like the stickleback too, he is a conscientious parent, instantly retrieving any eggs which fall to the floor of the tank and returning them to the nest, and keeping an eye on the youngsters for the first few days after they hatch out. It is a good idea to keep more than one female with each male as a solitary wife may literally be chivvied to death by over-amorous attentions. Feed as for guppies and swordtails, except that the fry should have egg-layers' Liquifry instead of live-bearers'.

The wild Fighting Fish is a dusky red in colour with green shadings, but selective breeding has now produced a variety of other hues, probably the most attractive and popular being the 'cornflower blue'. Whatever their colour, however, the male fighters are most handsome additions to any aquarium with their long flowing fins and tails, and even the more conventionally shaped female mirrors the colours of her spouse.

AMPHIBIANS

If you were to ask some of your pupils 'What is an amphibian?' the reply would almost invariably be 'An animal that can live on land and in water'. This is, however, not strictly true, for a hippopotamus lives both on land and in water and I feel sure that no-one would consider calling this huge mammal an amphibian.

The word 'amphibian' comes from the Greek and means literally 'double life'. The first part of an amphibian's life is spent in water, the place where it was born, breathing, like a fish, through gills; the second part may be spent on land breathing through lungs. (This is correct in theory anyway, but in practice there are many species that spend most of their lives in water, and a few that rarely if ever leave it, some even retaining their larval form and continuing to breathe through gills—e.g., axolotl, Japanese Salamander, etc.). For the purposes of this book we may say that there are four types of amphibian known to science: frogs, toads, newts, and salamanders; strictly speaking only the '*Rana*' and '*Bufo*' species are classed as true frogs and toads, the remainder of these first two groups being classed as 'tailless amphibians'.

FIRE SALAMANDER (Salamandra salamandra)

This intriguing little creature, dressed in a black skin with yellow spots and blotches, is an excellent example of warning colouration, the opposite of the camouflage displayed by many animals. Beneath two large yellow patches on the neck are situated poison glands which secrete a mild venom to cover the animal's body. Do not, however, be put off by the word 'poison'. Remember that there are various degrees of poison; the sting of a nettle is poisonous, for example, but, though uncomfortable, will have no serious effects on you. Similarly this salamander is a mildly poisonous animal—but only to the extent that it gives any predator that picks it up a very unpleasant taste in the mouth, and, if swallowed, possibly an attack of vomitting.

The Fire Salamanders derive their name from the belief held hundreds of years ago that they are born in fire. This is, of course, complete nonsense, for like most amphibians they are born in water. During the breeding season both male and female, normally mainly terrestrial, return to the water for mating. The male entices the female close to him, and then releases spermatophores (little sacs containing seminary fluid) which are collected by the female with her cloacal lips; however, if conditions are in any way not to her liking she can delay fertilization of the ova indefinitely. The eggs of this species, unlike those of most amphibians, develop inside the mother's body until the tiny transparent tadpoles are ready to be born alive; they already have minute limbs at birth, but as they grow the shape and colour of their bodies gradually change until after about nine months the little salamanders are ready to leave the water. Although they are now breathing through lungs, they must still keep their bodies damp because they breathe also through their skins, like all amphibians.

These salamanders make very good pets for the classroom as they usually feed very well and live very long lives—thirty years or more is not uncommon for this creature. Do not handle them more often than necessary as the warmth of the hand very quickly dries up the salamander's skin and makes him feel very uncomfortable.

When preparing a home for a salamander you will require an aquarium tank 30 cm–60 cm long, a quantity of garden peat, about two pounds of medium grade gravel, some sphagnum moss, and a few pieces of cork bark for it to hide under. Mix the gravel and peat together and saturate with water to make the environment pretty swampy, but leave a corner reasonably dry so that if the salamander feels he is getting *too* wet he is able to rectify the matter. Decorate the tank with the cork bark and moss, and also place a couple of small rocks in the aquarium. If you wish to attempt to breed these creatures a small pond will be necessary, and a large tank would be an advantage to avoid over-crowding.

AXOLOTL (Ambystoma mexicanum)

The axolotl, which lives in Mexico, is a very strange member of the amphibian family, a creature which remains in the salamander-tadpole or larva stage throughout his life, being incapable of turning into a proper salamander under natural conditions as the area that he inhabits is far too dry to sustain amphibian life on land. He therefore cannot leave the water, becoming sexually mature and breeding in larval form.

There are two main types or colour phases of the axolotl, black and albino, the latter white variety being one of the very few truly albino animals to be found in a natural state. They provide excellent examples of animal regeneration, quickly and readily regrowing the limbs and gills all too often bitten off by their voracious relatives.

No very special requirements are necessary to keep axolotls, in fact an aquarium tank with gravel on the bottom and a few rocks and water plants kept only at room temperature are all that is needed. They are carnivorous animals, and will simply need a few earthworms, slugs, insects, etc., on which to feed.

In confinement you can turn them into Tiger Salamanders (of which they are the larval form) if you so wish by reducing the level of their water 6 mm or so each day; if you do this, however, you will not be able to breed them as the metamorphosis makes them sterile.

The axolotl reaches a length of about 20 cm.

PLEURODELE NEWT (Pleurodeles waltl)

Newts and salamanders are very closely related, the most noticeable difference between them being that newts, generally speaking, tend to spend far more of their time in water than do salamanders. We do not here recommend anyone to keep British newts owing to the very precarious situation of the whole of our native herpetofauna. Instead we are going to examine a few of the far less rare newts from other European countries that one might keep in the classroom, more often than not very successfully.

The Pleurodele Newt is a creature which can grow to about 25 cm long (although it very seldom, if ever, reaches this size in a captive state), and is sometimes referred to as the Ribbed Newt. This alternative name is quite understandable if you have ever seen one of these mildly grotesque-looking animals, the ends of whose ribs are pointed and sometimes pierce right through the skin, possibly performing a defensive function by making the creature very unpalatable to any other animal who finds him on the bill of fare. Pleurodele Newts emanate from the Iberian Peninsula where they can be found living mainly in small pools. On the whole these animals do well in captivity, and can often be encouraged to breed—the female lays her eggs in clusters on small rocks and plants. Colouring is usually dark green with faded yellow, brown, and black blotches covering the body, and one of the most interesting features about this species is, I find, the oar-like tail.

The accommodation should consist of an aquarium tank approximately 60 cm long, lined with gravel, broken plant pots, etc., a few rocks and small stones, and a dry area to allow the newts to emerge completely from the water from time to time, although even this 'dry area' should always be kept moist as newts soon shrivel up if allowed to stay dry for even a relatively short period. Tilting the tank slightly at one end (the 'dry' end) will provide a nice little pool on the one hand and pleasantly moist 'land' on the other. Newts are excellent escapologists, and a firmly fitting top is a must if you wish to retain your specimens and try to beat the longevity record of about twenty years.

ALPINE NEWT (Triturus alpestris)

The Alpine Newt is very much smaller than the Pleurodele, and is to be found in the French Alps and Central Spain. This little creature is about the size of our Crested Newt, and, unlike the Pleurodele, is a very attractive amphibian, sporting a bright orange or red underside. I have heard of axolotl-like behaviour in Alpine Newts which have been born at high altitudes, inasmuch as late developers will remain in the larval stage throughout the winter, even to the extent of becoming sexually mature in this form, and remain in water during the cold winter months when normally they would seek out a damp place on land in which to hibernate.

House as for the Pleurodele Newt, except that a 'dry' area is not absolutely essential provided that there are a few rocks on to which they may climb to get out of the water if they should so wish. Both the Pleurodele and Alpine Newts should be fed on a diet composed basically of earthworms, but which can also include such items as woodlice, beetles, flies, and other types of invertebrate life.

MARITIME TOAD (Bufo marinus)

The Maritime Toad, a most fascinating creature found in South America and Africa, is similar to our own Common Toad, (*Bufo bufo*), except that it is much larger.

When talking about frogs or toads one is invariably asked: 'How do you tell a frog from a toad?', and this is a question which must be answered immediately to avoid confusion. As far as the spawn is concerned one does not and should not need to touch it in

Frogs	Toads
diurnal	*nocturnal*
jump	*walk*
smooth skin	*'warty' skin*
no poison glands	*poison glands (mainly behind head but also all over the body)*
	Said to be more intelligent

order to differentiate between frogs' and toads' because, as frog spawn is laid in a big mass and toad spawn is laid in strips or rows, a mere glance will tell one the answer. When they develop they are also different in many ways:

Toads have the most beautiful eyes, almost like jewels, which probably gave rise to the belief that they had precious stones buried in their heads—cf., Shakespeare's

> Which like the toad, ugly and venomous,
> Wears yet a precious jewel in his head.

Maritime Toads live very long lives, thirty years or more being not uncommon, and grow to such a size that they can cover a tea-plate. They are not difficult to keep in confinement, the most important requirements being warmth and dampness; a tank very close to a radiator would be ideal. They should be provided with a large dish in which to soak themselves as they absorb water through their skins, swelling up to gigantic proportions. This water they often discharge when picked up—a rather unpleasant defensive mechanism that makes you think they are urinating. Their tank will require a peaty base, kept nice and damp of course, and they will live quite happily fed on a diet of earthworms, beetles and other similar types of foodstuff; large specimens will even eat mice. Like the axolotl, they are very voracious creatures and must be given live food as generally they will only eat things which they can see moving. They seize the food either with mouth or with front limbs, and use their 'hands' to stuff in any bits and pieces left sticking out of their mouths.

GREEN TREE FROG (Hyla arborea)

Compared in size with the Maritime Toad, the Green Tree Frog is small indeed, measuring only 4 cm to 5 cm in length; however, what he lacks in size he makes up for in attractiveness, for this frog is the most brilliant green. He is to be found in many parts of southern Europe, being probably best known for his ability to walk up vertical surfaces, even glass, with the help of the sucker pads on his feet. Apart from being very beautiful animals, they are also accurate barometers, being extremely sensitive to barometric pressure and foretelling rain by croaking very loudly with voices quite out of proportion to their small bodies.

There are a great many species of frog living in most countries throughout the world, many with very strange peculiarities. Some species lay their spawn on the leaves of trees overhanging water, so that when the tadpoles hatch they fall off into the pond or stream below, while other species cannot even swim in the adult form, and would inevitably drown should they accidentally tumble into water.

In some country districts even today there is a belief that young frogs kill off the old ones by strangling them, but this is not true. The males are usually smaller than the females, and when the male thinks the female is nearly ready to lay her eggs, he will latch onto her back by grasping her firmly round the neck, often for a week or more, so that he is on the spot ready to fertilize the eggs immediately they are laid. Therefore the male is taken for the young frog killing the larger female or 'old frog'.

In captivity the Green Tree Frog must be kept reasonably warm, and is sometimes a little difficult to feed, especially during the winter, as he eats greenfly and similar small insects. They can, however, be hibernated, and this I think is the best course in view of the obvious difficulties in obtaining food. They will normally hibernate in cracks and crevices, but anywhere in a cool but frost-free room will suffice as long as they are left undisturbed. Unfortunately these frogs are fairly short-lived.

REPTILES

The word 'reptile', which incidentally comes from the latin verb *repere* meaning 'to crawl', has been known to make many people shudder. I think this is possibly due to the lack of understanding that most people have of these animals whose history is long and interesting. We must travel many millions of years back in time to meet the first reptiles that lived on the earth; these were the dinosaurs, some of which grew to about 27 m in length and weighed over 1000 kg. These great reptiles lived in many parts of the world; some were carnivorous, feeding on the flesh of other animals, some herbivorous, feeding on vegetable matter, and others omnivorous, eating both animals and plants. It is not my intention to say very much about dinosaurs—I mention them only to give you an idea of reptile ancestry.

Before we discuss the types of reptile that one might be able to keep in a school, I think it important to find out just what a reptile is. Primarily it is an animal that has no fur, hair, or feathers on its body, which is covered instead with scales of hard thick skin, as in the case of crocodiles, alligators, tortoises, turtles, terrapins and some lizards. The scales on the body of a snake, however, take the form of folds of skin which unfold when the creature swallows a large meal, allowing this to pass along the body to the stomach.

Reptiles are also cold-blooded animals, which means that they do not possess a body temperature of their own as we do, but adopt the temperature of their immediate surroundings. For example, if the temperature of the tank they are in is 27°C, their body temperature will be the same. Not all reptiles lay eggs (oviparous); some give birth to their young alive from their bodies (viviparous) in a manner similar to some mammals; and others are what we call ovo-viviparous—a little tongue-twister which means that the eggs are not laid but retained in the mother's body where the young hatch, arriving alive out of the cloaca (anal vent) to give the appearance of a viviparous birth.

People have on the whole been very biased in their regard for these unusual and uncommonly beautiful creatures, many saying that they hate snakes, for instance, when they do not know a single thing about them. Slimy, cold, poisonous, and creepy—these are just a few of the more common terms used to describe snakes by people with this 'snake-phobia'. Let us take each of these terms one by one and get a few facts straight. No terrestrial snakes are slimy, but they have a glossy sheen on their skins which is often mistaken by the uninformed for slime. A snake's skin is glossy in the same way that an ordinary domestic cat's fur is glossy, and it shows that the animal is enjoying good health. The skin will often lose this lovely sheen just before the snake casts or 'sloughs' (pronounced 'sluffs') it. When we refer to coldness we must remember one important thing—that a snake, being a reptile, is cold-blooded and if kept in a cold room will, of course, feel cold to the touch. As for snakes being poisonous, certainly some are, but out of about 2700 different species only approximately 150 are sufficiently so to harm a normal healthy human adult, thus leaving by far the majority completely harmless.

Snakes have no limbs, therefore they are unable to walk across the ground as you or I might, but if you examine the underside of a snake you will see the ventral scales, or 'scuts' as they are sometimes called, which are flaps of toughened skin fixed horizontally across the body. The job of these ventral scales is to mould the underbody to the ground, thus providing leverage for the snake's ribs. The ribs are not held together in the front by a sternum, or breastbone, so when the snake wants to move the ribs act as legs, and perform this function very efficiently too. If the snake wishes to go forwards, the muscles attached to the ribs pull them back, the ventral scales catch in the ground, and the snake moves forward; if it wishes to reverse, then the process is reversed.

Snakes are among the cleanest of all animals to keep in captivity, and from them you will certainly get no smell. They are also born escape artists, and a great deal of thought and care is required to provide them with an escape-proof enclosure. An aquarium tank that is being used to house a snake or any other type of animal that usually lives on dry land undergoes a slight change of name, and is known as a 'vivarium'. If you want to make your display look really attractive, buy a proper reptile vivarium, the design of which is quite different from that of an ordinary fish-type aquarium. The size of vivarium that you use to house your snake is very important as even these animals like a little room to be able to move about, but one 60cm long is quite sufficient for a snake of up to about 75cm long. Most shop-bought vivariums have an all-metal frame with glass in the front, back, sides, and floor, and a top usually of perforated metal. The front glass panel is removable for easy access to the inside, and is held down by its own weight so that locks or other fastenings are unnecessary. If some of the older boys are reasonably good at woodwork, they could well make vivariums in school, substituting wood for the metal and most of the glass—in fact, you really require only a glass front and perforated metal top.

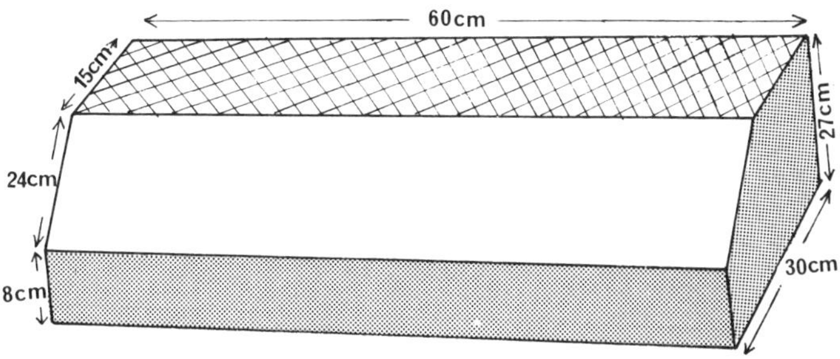

Typical Reptile Vivarium

I have found that heating a vivarium from the inside can present quite a headache, and is generally unsuccessful because snakes do tend to 'hug' warmth, and if you put an uncovered heater inside the tank the snake will often coil around it and burn itself quite badly. If the heater is covered, the warmth does not circulate properly and is difficult to regulate. By far the best method I have found is to raise the vivarium on to some kind of base and heat from below. Be careful that the heater does not actually touch the bottom of the tank, and place it off-centre so that there is more warmth at one end than the other; the snake can then choose its own degree of heat. The large vivariums in the reptile house at the London Zoo are heated by this method, elements being placed in the floor so that the snakes are provided with warm and cool places on which to rest. I normally use tubular heaters for small tanks (but the design is relatively unimportant); these are connected to thermostats which should be set to cut out when the temperature inside the vivarium reaches the required level.

For some strange reason few reptile vivariums are manufactured with provision for light fittings; if you buy one of them you will have to cut a small hole in the top and fit a bayonet plug or small fluorescent fitting—low wattage only required—to provide extra light.

Fresh water daily is a must as you will find that your snake will often use its water dish as a toilet—possibly thinking that the faeces will be washed downstream. The water dish should be large enough to act also as a bath and allow the immersion of the whole body. Food should be given once or twice a week as snakes usually eat as much as they can in one meal which lasts them for several days, sometimes even weeks. You may find that your snake will sometimes go on hunger-strike, especially when introduced to his new home for the first time; this is a characteristic of most reptiles, who like to feel really at home before accepting food.

GARTER SNAKE (Thamnophis sp.)

Garter Snakes are among America's most popular pet snakes and are found all over North America. They are very much like our own Grass Snakes in their habits and size. There are in fact eight types of Garter Snake, but the one that most frequently finds its way into both our pet shops and schools is the Eastern or Common Garter. I think that one of the most impressive things about the Garter Snake is the variety of its diet, which consists of frogs, toads, small lizards, earthworms and fish. This little snake is, of course, non-poisonous and, incidentally, also a non-constricting snake; it has to overpower its victims physically, using only the strength of its apparently feeble body. Garters tend to be rather more aquatic than Grass Snakes and this should be taken into account when preparing their captive home; I usually divide an aquarium tank in half, one half being dry land and the other water. One method that I tried did not prove too successful until modified; I placed a wooden barrier across the width of the tank and landscaped the section to be used for terra firma before pouring about 3 cm of water into the other section. The reason that this turned out to be rather a flop was that the water seeped through into the first section and we then had a pond and a swamp which was, of course, no good at all. However, I corrected matters by tilting the tank slightly so that the water was retained in its proper half.

Garters are often rather nervous when first imported, and I have found that they take a week or two to settle down. Do not allow pupils to handle a newly acquired specimen too frequently at first, therefore, bearing in mind that everything will be very new to it. Forcing one's attentions on a new arrival will only make it more nervous than it may already be. When handling it, always offer equal support to its whole body by placing one hand near its neck and the other about two thirds down its length; you will probably find that the snake will coil its tail round your wrist to give itself a little more confidence that it will not fall. Also never squeeze its body tightly (an important point to impress upon the children), as this is most uncomfortable for the creature and could possibly do it harm. As you all gain confidence in handling your snake, so it in turn will feel more confident in the hands of its new owners.

KING SNAKE (Lampropeltis getulus)

King Snakes come from the USA and are imported into this country throughout the warm summer months. There are several species of King Snake, including the California King, Florida King and Coral King, and they range from the southern part of New Jersey down to Florida.

The King Snake derives its name from its ability to overpower its prey, which consists of rats, mice, lizards and other snakes. I have often found King Snakes a little bad-tempered until they grow used to their new home and owner, and they may bite with little provocation during the first few weeks of captivity; as they are quite harmless, however, this need not be a matter for concern for the bite received from a non-venomous snake is unlikely to be as painful as one from a Golden Hamster!

The Florida King is the most frequently imported member of this family, and grows to between two and three metres long. Black in colour, with bright yellow bands the whole length of its body, this handsome fellow eats well and becomes quite friendly. A temperature of 21°C is quite sufficient for any of the King Snakes, and a covering of coarse sand and gravel on the floor of the vivarium will make them feel more at home.

BOA CONSTRICTOR (Constrictor constrictor)

Many tropical snakes do well in confinement, but one important thing to remember is that they require constant heat, and many of you will find that your school electricity account is a little higher than normal after installing such an animal.

The most common tropical snake to be found in English pet shops is the Boa Constrictor, or, to give the animal its proper name, the Common Boa, which comes from South America. Do not be put off by this animal's unfounded reputation; it is not the massive, fearsome creature that goes around crushing people to death that it is so often taken for. Docile by nature, and growing to a maximum length of about 4 – 5 metres in its wild state, the boa makes a lovely pet, ideal for children to handle. Rather more sophisticated equipment is required to keep this sort of animal than for the Garter Snake, and this, of course, is going to be reasonably expensive. A young boa would cost you between £6 and £8.50p before you even think about its housing requirements.

In its native South America the boa is the second largest snake —the anaconda, a water python recorded at a length of 9 metres, holding the record of longest snake for that particular country. These huge constricting snakes belong to the Boidae group, and are really the articulated lorries of the snake world, moving about very slowly because of their great bulk. This group is headed by the Reticulated Python of Malaya, the largest snake in the world, recorded at 11 metres long. One might well regard the Boa Constrictor as being a relatively small species when compared with this great monster.

Boas spend some of their time in trees and could be considered partially arboreal (tree-dwelling) animals. They feed on small warm-blooded creatures such as rats, mice and birds. This type of snake is viviparous (as described at the beginning of this section) the young being approximately 25cm long when born. From the moment of birth they must fend for themselves or die, and are, incidentally, perfect miniatures of the adult boa. I think that the most impressive feature of this species is its colouration—shades of black, brown, cream, grey, red and white forming a pattern which acts as camouflage, breaking up the outline of the animal's body and making it far more difficult for either prey or predator to see. Baos are principally nocturnal animals, hunting mainly small rodents which are more active at night. The eye of this creature is typical of nocturnal reptiles, the pupil being elliptical in shape. The prey is caught in the snake's jaws, where rows of long, sharp, backward-sloping teeth hold the victim in a vice-like grip while its captor wraps several coils around its body. Constricting snakes very rarely crush their unfortunate prey to a pulp, as many people imagine, because a clean kill is most important to them. The object is to kill the victim by applying just enough pressure to stop the animal from breathing and thus cause its death by suffocation; if bones are broken and protruding through the skin, the snake could be seriously injured when the meal is consumed. There are no incisor or molar teeth in a snake's mouth, so it is unable to bite pieces off or chew its food and has to swallow it whole.

Boas are rarely a problem to feed in captivity and specimens that I have kept or known personally have been quite greedy characters. The basic equipment required to keep a boa is as follows: one vivarium (dimensions depending on size of specimen) complete with a tubular heater, thermostat and light fitting. Only simple furnishings are required for the boa, and consist of a gravel covered floor, a branch for the snake to climb on, and a water dish large enough for it to soak itself in. Feeding is quite easy; simply give mice or rats depending on the size of your specimen once or twice a week, and always feed dead food if your snake will accept it, although many will not. Do not be too much put off by the length your boa might attain, as snakes rarely reach their maximum size in captivity, and should grow no more than approximately 23 cm per year. You would certainly have no trouble in selling the snake should he grow too big. A 2.5 m specimen that I kept for a number of years outgrew his vivarium and lived at liberty in one room of our house, much to the dismay of casual visitors who were not forewarned! I am not, however, suggesting that you give your snake the same sort of liberty in your classroom.

GREEN LIZARD (Lacerta viridis) and *WALL LIZARD (Lacerta muralis)*

There is a wealth of different lizards to choose from if you are contemplating keeping one in the classroom. In many sizes, shapes and colours, these creatures are, of course, the descendants of the giant lizards, the dinosaurs, long extinct, which wandered about our earth way back in the Mesozoic era. The two species which I mention in this section have been chosen because they are quite hardy creatures, fairly easy to house and look after, and require no very special treatment.

The Green and Wall Lizards are very widely distributed over Southern Europe, and provide excellent examples of sex-dimorphism (notable differences in colour, body-patterns, size, etc., between the sexes), the males of both species being more brightly coloured than the females. Both sexes of the Green Lizard are an attractive grass-green, but the male is distinguished by bright blue cheek patches during the breeding season, while in the case of the Wall Lizard the male has rows of light and dark brown spots along his flanks and a broad strip of vivid emerald green down the middle of his back, whereas the female is brownish all over with less patterning. Both these animals are tail-casters, that is they have the ability to shed their tails in emergency, so great care must be taken when handling them. If you do cause one to cast his tail he will usually regenerate a new one, but this never looks quite as good as the original.

A vivarium is the best place to house them, although I have heard of specimens escaping from captivity and living at liberty in this country in a wild state, so an *escape-proof* outside reptilium is also a possibility. Therefore these hardy little creatures do not need any special heating arrangements, although, like all reptiles, they do like to be able to bask in the sunshine—direct sunlight. At the same time a place to shelter from excessive heat is also important. The floor should be lined with sand and gravel, and plenty of rocks—and a few pieces of bark should be provided as they do like to be able to hide in nooks and crannies. Feed upon any form of insect life, mealworms and maggots being a good standby as they are usually obtainable from pet and fishing shops. I have found spiders, beetles and woodlice generally most acceptable to insect-eaters.

Length of life for the Wall Lizard is normally from two to three years, and for the Green Lizard anything up to five years.

TORTOISE

Each year literally hundreds of thousands of tortoises are imported into this country under the most appalling conditions, stacked one on top of another often for several weeks, and it is amazing that some of these pathetic creatures actually survive. The two main species which come in for this sort of treatment are the Spur-thighed or Greek Tortoise (*Testudo graeca*) and the Hermann's Tortoise (*Testudo hermanni*). Many arrive dead or dying, battered and crushed, and the shells of many of the remainder are so badly split that flies have been laying eggs in the animals' wounds and they have to be put to sleep—and still, in spite of all this, they continue to arrive throughout the summer months. Fortunately some of the survivors fall into the hands of people who know and understand these reptiles, but I would estimate that less than five per cent grow to maturity. So if you are contemplating keeping a tortoise it is only fair to warn you that it may not live very long unless you and your pupils are prepared to tend it as an animal close to extinction, for in the very near future this is just what it will be.

The most noticeable thing about a tortoise is, of course, its shell; try asking your pupils what they think this shell is made of and the chances are that many will think it is wood, or stone, brick, ivory, or plastic—you will be very lucky if you receive the correct answer straight away. The answer is bone—bone and skin—the same sort of stuff that their bodies are made of, for that is what the tortoise's shell is—its body. The top shell, or carapace, is fused to its vertebrae and is covered with hard, dry plates of skin. The smaller bottom shell, or plastron, is fused to the carapace, and affords the animal a great deal of protection from prospective enemies. Four legs, a head, and a tail protude from this creature's shell-like body, and these can be withdrawn into the shell should danger approach. The tortoise's two front limbs act rather like a front door and close before the animal's head making its defences almost impenetrable. You will notice the stubby claws on each of the four legs, which are used to dig into the ground and enable the heavy tortoise to walk along, sometimes at surprising speed.

Throughout the summer months tortoises must constantly be feeding, for it is most important that they build up large reserves of fat to use during their winter hibernation. They must also receive plenty of sunlight if they are to remain healthy so they should be allowed out of doors as much as possible, although shelter should always be provided to protect them both from inclement weather and too great heat in the unlikely event of their having too much sunshine in England! Tortoises are herbivorous and will eat cabbage, lettuce, carrot tops, clover, dandelions, tomatoes, cucumber and, from time to time, fruit such as strawberries, blackberries, apple and banana. They are often also fond of virginia creeper leaves, and once they have started feeding they appear to forget everything else and just concentrate on food.

As the cold weather approaches the tortoise will start to look for a place in which to hibernate, and this is the time to bring it indoors—but only into a warm classroom if you do not intend to let it hibernate. Personally I do not hibernate my own tortoises because so few come out of hibernation the first year. I prefer to keep mine in a semi-warm room where they will remain torpid and eat seldom; many people disagree with this, but so far I have kept them all *alive*. If you do decide to hibernate them, a cool stock cupboard or an outbuilding away from frost is the best place. Put plenty of straw in a box and place the tortoise in the middle of it so that it is well insulated against the cold. It should quickly settle down and sleep through the winter to awaken at the onset of spring.

It appears that few people are able to differentiate between the sexes of tortoises, but this is a relatively easy task. If you examine the plastron you may find a shallow denting. If there is no dent, then the tortoise is probably female, but if the plastron is concave then it is almost certainly male.

One word on handling—always place the fingers around the plastron with the thumbs on the carapace, and, when lifting, remember it may be heavier than you think, so do please be careful as, if you or any of the children drop a tortoise, this will almost certainly damage its shell and may even cost it its life. If an accident does occur and the shell *is* split, a minor crack will often heal if painted with a little Friar's Balsam, but more serious damage requires the attention of a veterinary surgeon; in any event, do remember to ensure that flies are unable to get at any wound that a tortoise might receive.

TERRAPIN

Terrapins are, as one might expect, very close relatives of the tortoises—one might describe them as their aquatic cousins. Many of them are imported into this country throughout the spring and summer. Most often to be seen in pet shops is the Red-eared Terrapin (*Pseudemys scripta elegans*), which often arrives as a tiny hatchling about 2½ cm long. Like the two species of tortoise previously described, very few survive in captivity and so many have been imported that reports now come to my ears that they are becoming much more difficult to find, a sign which means 'don't buy them, leave them alone'. The principle of supply and demand can very easily be reversed—if you do not buy them, the pet shops will not be able to sell them and the suppliers in turn will not be required to continue to supply them. If you want to keep a terrapin, might I suggest the European Pond Tortoise (*Emys orbicularis*) or the Spanish Terrapin (*Clemys leprosa*), as both these species are imported in much smaller numbers, and this will give at least a few of the red-ears a chance to increase their numbers. The European Pond Tortoise is a very hardy species, quite capable of living out-of-doors all the year round if given facilities for hibernation, but the Spanish Terrapin is slightly more delicate and should be kept indoors at room temperature at least during the winter. If you intend to provide both species with outside accommodation during the summer months a small pond is ideal, then they will be able to live their lives quite naturally and only a low wall or fence is necessary to keep them from wandering too far away from the water.

One of the biggest problems suffered by people who keep terrapins is that quite often, due to incorrect feeding, the creatures fall a prey to soft-shell, a complaint which literally is what it says as the shell does indeed become soft. Very little work is required to keep these animals healthy, but what does need to be provided must be properly provided. There are four main prerequisites for long and healthy life: correct feeding, sunlight, water and dry land.

Correct feeding, whether it be of terrapins or any other species of animal is obviously of great importance. If you feed your terrapins on best steak you may feel that they are getting the best possible food that money can buy, but you might just as well purchase scrag-end, for a regular diet of either of these items without the necessary roughage will produce all sorts of dietary deficiencies. Calcium is most important but, incredibly, the very thing which is most often forgotten. Feed whole organisms whenever possible; chopped up dead rats and mice are excellent, together with earthworms, slugs, snails, etc., plus additional calcium if required. Lack of sunlight is also a major cause of soft-shell; direct sunlight is best as glass filters out the ultra-violet rays. The water need not be heated for either of these animals, but if you intend to allow the European Pond Tortoise to hibernate in the winter, you will need to provide quite a reasonable depth of mud at one end of the pond so that he can dig himself deep into it below the frost line to spend the winter months. The Spanish Terrapin could share the same accommodation during the summer, but must be brought indoors during the winter and kept at room temperature. It is also very useful to add a little limestone to the water as this will help to prevent soft-shell.

Remember that, whether you keep terrapins inside or outside, they require land as well as water, and a reasonable amount of land at that. I have seen too many schools keeping terrapins in nice large aquarium tanks with plenty of water but hardly any dry land at all. It is important that terrapins and turtles should be able to bask on dry land if they so wish, and this will also help to keep their shells nice and firm. If an outside pond is out of the question (although this is undoubtedly the best way of keeping these two species of terrapin), then an aquarium tank will suffice. Both these species grow to a length of approximately 25 cm and an animal this size is obviously going to require sizeable accommodation, so I should be inclined to choose a specimen ten or fifteen centimetres long, for which you will require an aquarium about 1 m × 30 cm × 25 cm. Do not line the bottom of the tank with gravel as so many people do because this only encourages sores on the plastron as well as harbouring scraps of uneaten food; in fact there is no need to line the tank with anything. I have found that building bricks make very good dry land areas, and above the land should be placed an infra-red bulb for the creatures to bask under. Do not hang this too low, however, as these bulbs are very powerful and can very easily give the terrapin heat-stroke as well as actually damaging the carapace. No heating of the water is required for either species in summer or winter.

Whether you keep your terrapins outside or inside, do not attempt to put them in with pet fish, tadpoles, frogs, etc., or you will find out that terrapins are also very fond of nice fresh live food!

BIRDS

Very few schools at present seem to keep birds, and this, I think, is due to the fact that most people believe that the only method of keeping them is to confine them in a small cage. This, however, could not be further from the truth, although many—to my mind, far *too* many—people do keep birds in much too restrictive accommodation. The RSPCA requires that all birds kept in captivity are provided with enough room for them to stretch their wings, although strangely enough this does not appear to apply to parrots, often kept in 'parrot cages' and completely incapable of flexing their wing muscles to the full extent. In any case, if there is room in either the entrance hall or in the playground, we recommend that all birds be kept in aviaries. In the next few pages we suggest a selection of birds some of which can be kept either inside or out, depending on choice.

BUDGERIGAR (Melopsittacus undulatus)

Budgerigars have been kept as pets in this country for many years and can be found living under a multitude of conditions—some in large aviaries set in the heart of the country, others solitary specimens enduring a lonely life in drawing-room cages at the top of sky-scraper blocks of flats. Originating from the Grass Parakeet of the grasslands of Australia's outback, and originally green in colour, the budgerigar that we see today has come a long way. Being greatly favoured by bird breeders because they are easy both to keep and breed, 'budgies' are found in many different colours developed through selective breeding—lutino (yellow), violet, blue, albino, cinnamon, pied, opaline and other colour deviations.

Sexing adult birds is not too difficult a task as the males have a blue 'cere' (the 'bobble' just above the beak, so called because it looks rather like a little dollop of sealing wax, *cera* being latin for wax), and the females brown. Male and female are, as one might expect, called cock and hen, but the newly-feathered babies are known as 'barr-heads' as the stripes on the head at this stage extend right from the nape of the neck to the cere; the ceres are always brownish in the youngsters, regardless of sex.

Feeding presents no problems as budgie seed is always available at pet shops either loose or in packets, and chickweed, shepherds' purse, iodine niblets, and ordinary greenstuff should be given from time to time. A cuttlefish bone should be placed in the cage or aviary to provide essential mineral requirements, and is normally suspended from the top of the cage or side of the aviary depending on the type of accommodation.

The trimming of claws and beak can be safely undertaken should it become necessary as long as proper clippers are used and the extra growth is taken off down to, but avoiding, the quick. It is advantageous to wrap glasspaper around the perches (perches can be bought already thus equipped) both in order to keep the claws in trim and to help to prevent a rather unpleasant condition called 'slip-claw'. Certainly the most common ailment in budgerigars is scaly-face, when the cere and much of the face become encrusted with a hard excrescence rather like scaly-leg in poultry, but this can be cured fairly easily by painting on various proprietory preparations available from most pet shops.

If you wish to try your luck at breeding budgerigars you must have a fairly even distribution of the sexes or fighting may well break out as budgies are usually monogamous. It is possible to breed these birds in small cages, but better results are likely in an outside aviary; in either case the same type of nest box will be required. This may be bought ready-made or would make an excellent exercise for the woodwork class; in the latter case, it is best to buy a 'concave'—the base of the nest box, which is a single piece of board with a circular dip in the centre to keep the eggs together. The box, which can be from about 20 cm to 25 cm high, can be easily constructed upon the ready-made base and should incorporate an entry hole with a short perch outside it, and a removable lid. When several nest boxes are required in a large aviary, they should always be placed on the same level to prevent the birds bickering over who should have the top flat. As budgies are likely to wear themselves out by breeding all the year round, it is a good idea to introduce the nest boxes in the spring and remove them again at the onset of autumn. Up to eight eggs are usually laid, pure white in colour, and the babies are born blind, naked and quite helpless, and should be ringed before they are three days old. There are different coloured rings for each year which are obtainable from pet shops. Budgies, like all parrot-like birds, are what is called zygodactylous, that is they have two toes at the front of the foot and two at the back, unlike pigeons, blackbirds, sparrows, chickens, etc., which have holothecal feet with three toes in front and one at the rear. To place the ring on the leg you must press the hind toes back until they lie along the leg and slide it over them. The toes will return to position and thus keep it in place.

Length of life varies enormously; the average appears to be about twelve years but I have kept one for fourteen and have known others to live as long as seventeen years.

CANARY (Serinus canarius)

Just as the budgerigar originated from a wild parakeet, the canary originated from a wild finch, the Serin or Singing Finch. Canaries are, of course, most famous for their song and probably achieved their first popularity during the Victorian era when many people kept singing birds for pure ornament rather than in any serious attempt at pet animal husbandry. Today many colour varieties are available to the pet keeper—yellow, buff, red, orange, apricot, bronze, opal, etc., and also many different types—such as Yorkshire, Norwich, Gloster, Lizard, Frill, etc. The various colours may be arrived at either through selective breeding or by feeding colour-impregnated food—for example, sprinkling cayenne pepper over the seed will produce red plumage.

Feeding presents no problem as plain canary seed is available from most pet shops and large departmental stores. As for breeding, nest pans about 12 cm in diameter made of either pottery or plastic, with drainage holes, together with balls or sheets of nesting material are all that are required and are also available from pet shops. The general breeding instructions for budgerigars are also applicable to canaries, although it is inadvisable to keep the two species in the same aviary as budgerigars often have a rather unpleasant habit of biting the legs off non-parrot-like birds.

DOVE AND PIGEON

There is sometimes a little confusion as to what a dove actually is. A dove is really no more than a small pigeon, and for the purpose of this book we are going to treat both doves and pigeons in a rather indeterminate sort of way because their needs are so much alike.

For aviary birds we have the choice of Java, Barbary, Turtle or Collared Doves and Wood Pigeons, and it must be stressed that these birds are not ideal for free flying as the last three are liable to depart to join their wild fellows while the first two are far too trusting and likely to walk up to the nearest cat with obvious dire results. For free-flying or homing pigeons you have plenty of choice, the most popular being the Fantail, which is generally white but comes in several other colours; there are also Modenas, Rollers, Tumblers, Tipplers, Pouters, Gimples—and, the biggest of all the domestic breeds, the Runts. Many of the domestic varieties are descendants of the Rock Dove, a bird very fond of nesting on rocky ledges, etc., hence the nesting habits of the pigeons in London's Trafalgar Square.

You will need to provide free-fliers with a dovecote or pigeon loft—a good idea for woodwork—which you can design yourselves. The landing platform should be covered in with wire-netting for at least a fortnight after the birds are introduced—to be on the safe side I usually leave mine confined for three or four weeks—so that they have a chance to see around them and get to know the lie of the land before they are released in the hope that they will return. They generally do!

Their diet consists mainly of grain, bread, greenstuff, etc., and they should be fed once a day. For breeding you will require a large nest pan approximately 20 cm–25 cm in diameter. It is very strange, but both doves and pigeons invariably lay two eggs in a clutch and the youngsters, which are called squabs (parents cock and hen), are always one male and one female. The youngsters are usually ready to leave the nest at about three weeks, but are completely incapable of flight until they have their full adult plumage. Until the squabs are fully able to fend for themselves the parents feed their offspring on a substance called 'pigeon milk', which is really partially digested food regurgitated into the throat and mouth, and the youngsters simply root around and help themselves. We do have at the time of writing a young pigeon squab who refuses all food unless it is first chewed up by us and he is allowed to ferret about for it with his head stuck into our mouths —a process not to be recommended, but often unpleasantly the only way to get a squab to feed.

If you really want to be different, you could keep such exotic varieties as Diamond Doves and Bleeding Heart Doves, which are quite easily obtained, although not particularly hardy, and should ideally be kept inside the building in a warm room.

BANTAM (Gallus gallus)

Bantams have been kept in a number of schools for many years with great success, so we have decided to include a few words about keeping them in this volume.

All the many varieties that we see today are descended from the Red Jungle Fowl of India, in fact the Rhode Island Red bantam looks very much like this species. Probably the most attractive is the Silky, a really beautiful bird with light fluffy feathers covering him literally from head to toe. These feathers appear in a number of colours, the most common being white, but there are also grey, black and a reddish brown. The males usually have very striking blue wattles and a rather broad comb. Silky crosses are reputed to make very good foster mothers, so if you have any pheasant eggs, etc., you want hatching, a broody Silky or Silky cross may well oblige.

Yokohamas are very much like the Jungle Fowl in appearance, except that the males can have tails up to 8 m long and would obviously present housing problems for a school—so it might be a good idea to give this breed a miss. Another very attractive variety is the Mille-fleurs, which literally means 'a thousand flowers', this name being derived from the myriad white specks all over the body. The Polish bantams provide yet another eye-catching variety, with puffy little top-knots of feathers adorning their heads —and, of course, all the usual breeds of normal poultry, such as Maran, Wyandotte, etc., are also represented among these miniature fowls.

Bantams are kept in just the same way as other domestic poultry and will require a fairly large run as they like to be able to scratch about and if kept in too small a run will soon turn it into a very messy mud bath. A house is also required for them to retire to at night and during inclement weather, and this must contain both a perch and nest boxes. It is a good idea to have the nest boxes attached to the outside of the house, with access from outside and inside to facilitate egg collection. The run and house should be cleaned out about once a week, and the latter sprayed with an insecticide (available in the pet shops) to prevent scaly-leg and red mite. The diet should basically be one of grain and mash, and must include crushed oyster shell and grit, offered separately, to ensure proper digestion of food and nice firm egg-shells.

Do not be alarmed if the egg that your bantam lays for the first time is very small, as the first egg is very often a trial run and can sometimes be as small as a blackbird's. A broody coop can be added alongside the house if you do not want the hens to become broody, consisting of a small hutch, raised on legs, with a wire mesh or slatted floor. If placed in such a structure a broody hen will normally lose all desire to 'sit' after two or three days, as the nature of the floor makes such an occupation very uncomfortable for her. If a hen becomes broody (a condition easily recognized by her unwillingness to come off the nest, indeed she will often peck you quite savagely if you try to dislodge her!), and you want to try your hand at hatching out some eggs, she must first be separated from the rest of the flock and placed in an enclosed box with plenty of straw. When she has had time to 'settle in', slip the eggs under her, preferably at night. Thereafter she must be let out once a day for exercise, food and drink. The incubation period is three weeks, and when the chicks hatch out they are covered in very soft fluffy down and should be fed upon chick corn or chick crumbs, available from most pet shops. You could, however, try hatching your eggs in an incubator. These are normally bought from electrical shops, although pet shops do occasionally keep them, and they cost between £3 and £6. After the chicks hatch you will need a warm shed in which to rear them, with a small, low wattage lantern for them to snuggle up to as, remember, there is no mother for them to cuddle under. I can well remember the occasion when I was presented with two cock domestic chicks, rescued from a factory farm in Kent. I took them back to my London home, and they spent their first night in my bedroom in a nice warm straw-filled cardboard box. I was rudely woken by one of the chicks sitting on my head, while the other was trying to snuggle between the sheets around my neck. I first removed the chick from my neck and put him down on the pillow and was just about to rid myself of the gentleman sitting on my head when—ugh!—what have I put my hand into?—and at three o'clock in the morning! Needless to say, the next night found them in more secure accommodation.

Incidentally, one final word on feeding; if you intend to keep bantams purely for egg production and feed them on layers' mash or layers' pellets you can literally double your egg output with no adverse effects to the birds.

DUCK AND GOOSE (Anas platyrhynchos and *Anser anser)*

Ducks and geese are very interesting creatures to have around the place and also a very useful source of a little extra money for dwindling school funds.

If you want to breed ducks for the table (and I suspect that you do not!) the lovely white Aylesbury is the best table bird; if you want eggs, the brown Khaki-Campbell is certainly the best egg-layer; if you simply want a decorative, but still domesticated, variety then the black and white Muscovy, a tree duck, could be the answer to your needs. Muscovies, however, being tree ducks, are very good fliers, and you have to clip back the primary feathers of *one wing only* to prevent them making their escape. If you clip both wings, they will soon manage to recapture the art of flying with shortened wings, and the chase will be on; clipping a single wing simply makes them rather off-balance if they try to take off and this ensures that they will have to stay around. Other more exotic varieties that are also quite easy to keep are Carolinas, Mandarins, Shelducks, etc., and all are kept in much the same way as domestic poultry, although they are sometimes a little troublesome and noisy when you come to put them to bed for the night. The females, however, lay their eggs by night, unlike the bantams who lay theirs by day, and you can therefore collect the eggs when you let them out in the morning. They prefer a much more sloppy mash than the other poultry, with the addition of extra protein in the form of fish meal or something of that kind, and also require greenstuff to peck at when they feel like it.

A pond is really essential if they are to be kept completely happy and contented, as they like to be able to 'poddle' (pond-dipping!) as well as swim around. You can buy glass-fibre ponds from gardening suppliers if you are unable or would prefer not to make your own. These come in a variety of shapes, sizes and colours. Apart from the educational value of having ducks about the place, imagine watching your own ducks on your own pond in your own school—a really magnificent sight. One very important thing concerning duck eggs is that they must be kept very clean as the shells are hydroscopic, which means that they can absorb any moisture complete with attendant dirt, so a good wash as soon as they have been collected is an excellent idea.

On the whole domestic ducks do not make very good mothers, with the exception of the Muscovy, and, if allowed complete freedom night and day, tend to drop their eggs wherever the fancy takes them, apparently not in the least concerned for the eggs' safety. If when you find them you want to hatch them, place half a dozen or so under a broody bantam, or more in an incubator, and you will stand a fair chance of success. It is also most amusing to watch the look of utter consternation on the face of the ducklings' foster mother when her progeny first take to the water, where she is, of course, quite unable to follow them! The incubation period for most ducks is about 28 days, and 31–32 days for the Muscovy. The Muscovy makes an excellent mother, though she prefers to choose her own nesting site, and is more likely to settle down under the shed than in your thoughtfully provided nesting boxes. Ducks make their nests from a mixture of their own down and hay, straw or grass, and turn up the edges of the nest to cover the eggs completely when they go off to feed—as do geese, who have similar requirements as far as husbandry is concerned. Feeding for both ducks and geese is much the same as for other poultry, though the mash should be mixed far more sloppily, as they do like to be able to 'slurp' up their food. Geese are similar to ducks in many ways, although, being grazing animals, they require a bigger area of grass if they are not to turn it into a mud-wallow; the main differences are that geese tend to be larger than ducks, are nocturnal rather than diurnal, and both sexes are coloured alike and share in rearing the young.

The most common domestic goose is the Roman, a descendant of the wild Greylag, which is usually white or grey, or a mixture of the two. These were used as watch-dogs by the Romans, because they make quite a racket at the approach of man or beast; the Chinese Goose, the next most popular domestic variety, descended from the Swan Goose, is even more efficient in this respect, being considerably noisier than the Roman Goose. The Chinese are very attractive animals, tall and slender, and coloured rather like seal-point Siamese cats with a white throat merging into a creamy breast. Geese, like ducks, feel much more at home with a pond to swim about in, and lay between four and five eggs in a clutch which, like ducks, they incubate for approximately 28 days. On the whole ducks and geese are far more intelligent than hens, and make very good pets.

MAMMALS

When one bears in mind that we are ourselves mammals, it is quite startling how many people, especially young children, have not the remotest idea what a mammal is, so perhaps a definition would not be out of place here. A mammal is a vertebrate—i.e., has an internal skeleton composed of bone; it is warm-blooded—i.e., it has a definite body temperature of its own, varying quite a lot from one species to another but reasonably constant within each species; its body is covered with hair, not feathers as in a bird, or scales as in a reptile or fish; with only two minor exceptions, it does not lay eggs but bears its young alive; and most important of all, in fact the very quality from which its name is derived, it feeds its young upon its own milk. The mammal is, of course, the highest development on the evolutionary tree, probably largely owing to the high degree of specialization it has attained in reproduction. Whereas among the invertebrates parental care is virtually nil and they have to lay hundreds or even thousands of eggs in order to ensure that just two survive to carry on the species into another generation, the mammal, the development of whose egg takes place within the safety of the mother's body, and whose young even after birth can be fed without her leaving them in her quest for food, can afford in many cases to produce only a single infant with a fair surety that it will live long enough to become itself a parent. The degree of specialization is quite incredible in its efficiency. Many mammals after giving birth devour the placenta, this containing various substances beneficial to the mother after her ordeal, while the first milk, the colostrum, contains a mild aperient designed to set the new-born baby's digestive tract in motion. Everything is catered for—not only is mammalian milk a steady supply of a nutrient ideally constituted for its purpose, but it is also capable of transmitting to a large extent the mother's acquired resistance to certain diseases. Yes, the mammal is a miracle of specialization and organization; what a pity we do not appreciate the amazing chemistry of our own bodies more—but perhaps we shall if we first learn what we really are!

GOLDEN HAMSTER (Mesocricetus auratus)

For many years now the Golden Hamster has been a great favourite in schools throughout the country, but how it has remained so is quite beyond me, although possibly until now there has been little choice in the way of small furry creatures that would settle down well in the classroom. The hamster's semi-nocturnal habits coupled with its quick-to-fire temper and not unpainful bite, often sparked off by the prodding fingers of well-meaning and enthusiastic youngsters, to my way of thinking makes it a very *un*suitable creature for the primary school, especially as it is often the first unusual animal that young children, particularly in large towns, are introduced to, and I mention it in these pages only because it is such a popular creature.

As a species the Golden Hamster is a charming little fellow, dressed in sandy-coloured fur and with large appealing eyes; no wonder so many people regard it with such great affection. It is, of course, a comparatively recent discovery, and scientific note was not taken of it until about 1930 when a zoologist called Professor Hindle found one mature female and eight immature young, from which all our present day pet specimens are descended. These animals do in fact originate from Israel. They are desert dwellers and dig tunnels in which they spend most of their lives far below the surface of the ground. Due to the sandy colour of their fur, which affords excellent camouflage, hamsters are rarely seen in their native country, and if you were to put your hamster in the sand pit you would notice how well it blends in with its surroundings. The tunnels that it digs are used as food stores as well as sleeping quarters because when there is cold weather about the sensible hamster prefers to be able to stay indoors. In actual fact hamsters do not spend more time than they have to in looking for food, and are very selective about the time of day when they do it. Under natural conditions, the hamster will venture out at dusk and in the early morning because it is a twilight or crepuscular animal. When looking for edible items the hamster, being herbivorous, is after anything 'planty'—roots, seeds, young shoots, leaves, berries—and fills its cheek pouches with anything that it finds to eat. The cheek pouches can be filled to the most ridiculous proportions, and even when the creature's head is almost enveloped in them it still tries to stuff in more. In captivity it likes to hide its food where only it knows where to find it, and if allowed to run around a room for any length of time with its pouches full it will find somewhere to conceal its hoard of goodies; you may come upon it under a carpet or beneath the cushion of a chair, in any little snicket it can find that it thinks you won't.

The hamster has very long incisor teeth which have no roots and so are constantly growing; it must therefore keep them at their proper length, and this entails quite a lot of gnawing. If it is housed in a wooden cage it will often eat its way out of it and, if not presented with anything to gnaw, will often be found chewing the wire mesh of its cage or simply grinding its teeth, a most unpleasant noise to listen to. Its home need not be an elaborate affair, but I do not recommend a wooden cage because this cannot be guaranteed to hold both the hamster and its gnawing little ways, and therefore a metal one is a far better idea. Many pet shops supply cages tailor-made (or so they think) for these little creatures, but please, please buy a large one as the standard small hamster cage is far *too* small, and restricted movement will not improve its temper.

If you wish to breed hamsters you will require only one male to several females, as hamsters are polygamous creatures, but be warned—they can have about six or seven litters per year with an average of five or six youngsters in each. Remember to remove the male after mating as the female is liable to murder him as the time for parturition approaches. Ideally hamsters should be housed on a sawdust or, failing that, wood shaving floor, as this will soak up urine, and they should be cleaned out regularly except, of course, when breeding is taking place, when they should be left undisturbed for at least ten days after the birth. Bedding should be changed about once a week, and may consist of cotton wool, wood shavings or straw—although probably the best of all is soft toilet paper which they will be happy to shred up themselves.

Feeding should present no problem as there are various types of hamster food on the market which are always easily obtainable. Do remember, though, to give fresh fruit and vegetables too, as a lot of the creature's body moisture is obtained through its food—lettuce, cabbage, carrot, apple, grapes, etc., and an occasional small piece of meat all go down well. Contrary to what many people believe, hamsters *do* require a drinking bottle with fresh water provided daily. Properly looked after, your hamster may live for up to three years, although the average is eighteen months to two years.

MONGOLIAN GERBIL (Meriones unguiculatus)

The Mongolian Gerbil is one of my favourite rodents; it has all the advantages of the Golden Hamster without many of the disadvantages. Like the hamster, the gerbil is quite a recent discovery whose existence was noted only in 1954 when a group of scientists from Yale University were working near the Mongolian plateau. They returned to their university with eight pairs and six spare males, and it is from these few that all pet Mongolian Gerbils we see today originate.

Gerbils are excellent creatures for schools to keep. They have the advantage over hamsters because they are diurnal, very placid, generally have smaller litters, and are very resistant to disease. I have handled literally hundreds of these creatures, and so far as I can remember have been bitten only once, and that was not in anger but more out of curiosity really—I think the animal wanted to find out just what human beings tasted like!

Housing a gerbil presents no problem, but, like hamsters, rats and mice they are necessarily very fond of gnawing, so a metal cage is again the answer if one is to retain one's pets. House them in the same sized cage as the hamster, with the same sort of flooring, bedding and foodstuffs.

Unlike hamsters, rats and mice, however, gerbils are monogamous, and breeding can sometimes present a problem because you have to find a compatible pair; if the male does not like the female of your choice it is not really any good leaving them together in the hope that they will grow to like one another; they will probably fight like the devil, so at the first signs of this, unless its just a lovers' tiff, of course, separate them straight away. Similarly, if one partner dies it is quite a problem replacing the deceased and it may be found, unless the surviving partner is very young, that this is virtually impossible.

Gerbils also appear actively to enjoy being handled, but try to avoid overdoing this for, like all other animals including the human animal, even gerbils have their 'off-days'. Do ensure that children do not disturb newly born young for at least three weeks, as gerbils develop much more slowly than rats and mice, and mother is likely to kill them if they are disturbed too early. There is no need to remove the father after the youngsters have been born, for both gerbils are usually very good parents.

RAT AND MOUSE (Rattus norvegicus and *Mus musculus)*

I know of quite a few schools which refuse to keep rats or mice on the premises because of their smell, and this, of course, is quite understandable; however, if these rodents are kept under proper conditions, fed with the right type of food and cleaned out regularly, this problem will not arise. Both rats and mice, in fact all rodents, have a certain muskiness about them, but it must be remembered that all species have their own personal smells, including the human animal. If you feed mice on cheese or chocolate these are the sorts of food that do make them smell, and a diet of this type will certainly kill them very quickly anyway.

The Brown Rat *(Rattus norvegicus)* is the wild ancestor of the domestic or pet rat. This animal, as his biological name suggests, originated in Norway, and has spread himself all round the world by stowing away in ships carrying any sort of foodstuff and thus being carried into the warehouses of foreign countries. As many of you I feel sure will realize, the Brown Rat has absolutely nothing to do with the plague, this being spread by the Black Rat (*Rattus rattus*), occasionally seen living in or very close to sea ports.

Brown Rats are extremely adaptable creatures, and can live under almost any conditions; they have even been found living in the refrigeration compartments of ships after growing longer fur coats to keep themselves warm. They are also highly intelligent, and their intelligence coupled with their adaptability has made them one of the world's biggest pests.

The pet rat comes in a variety of colours, the most common being pink-eyed white; black, brown, grey, red, chocolate, champagne and agouti are also found as 'self' or plain colours, and there are 'hooded' rats too which are white with coloured heads and dorsal stripes. Whatever their colour, they all make good and interesting pets. Rats, like gerbils, like to receive a lot of attention in the form of handling, but here again don't overdo this, and never allow a child to *grab* any animal. Careful and considerate handling can do nothing but good to the temperament of a rat that tends to be nervous.

Selecting a rat in good health is not too great a problem; look at its coat, eyes, ears, nose, limbs and shape of body. The coat should be nice and shiny, the eyes clear, the nose dry and active (sniffing), the limbs and body free from rickets and other apparent diet deficiency ailments. Many Education Authorities are now establishing biology centres from which schools may obtain the usual types of animals and plants used in biology courses, so if you have such a centre in your area you could contact the warden who would be pleased to give you further details. You could also contact a local secondary school to see if they have any surplus stock, in which case the head of the biology department might even let you have one or two free of charge. Failing these, there is always the local pet shop.

Housing and feeding are easily undertaken, but the cage must be reasonably strong, of thick timber with a mesh front—perhaps constructed by some of the older children, or of metal, or purpose-built of hard plastic. Supply several items to enable the rat to take exercise and amuse himself—e.g., exercise wheel, ball, cotton reels, etc. Rats will eat just about anything, so it is very important to watch their diet; feed on fresh vegetables, nuts, seeds, cornflakes, bread and a little meat, making sure that fresh water is always available.

If you intend to have two or more rats you could breed them quite easily, but the male should be removed when you think the female is pregnant as male rats are the natural enemies of the babies, in fact wild rats can sometimes be controlled if as many females as possible are killed but any males captured set free to hunt and kill the new-born young. The male should not be reintroduced to the female and her family for at least three weeks. I am sure there is no need for me to repeat the warning about children interfering with either the pregnant female or her brood.

Mice are often more popular than rats, and the little mouse (*Mus musculus*) can be seen in a myriad of colours, in fact almost any colour or combination of colours that you care to think of. Again the most common appears to be the albino (pink-eyed white) which is the direct descendant of the House Mouse. Mice which are white mottled with another colour are called 'broken mice', and a recent addition to the mouse world has been the hairless mouse, a creature completely devoid of any fur and one that I personally feel very sorry for because it looks so weak—although it can in fact be very hardy. A friend of mine had a pair which escaped from his private collection and about a month later disturbed a female which had given birth to a litter in an unheated garage in the middle of winter—and I believe that they all survived.

Housing and feeding are really the same as for rats and, as they are prolific breeders you are likely to end up with a lot of surplus stock. In this event remember that if you keep any carnivores whole organisms provide the best possible nourishment, and dead mice make excellent food.

DOMESTIC RABBIT (Oryctolagus cuniculus)

Breeding chart for rats and mice

	Rats	Mice
gestation period	28 days	21 days
eyes closed	10-12 days	10 days
wean	14–22 days	14–21 days
reach maturity	2–3 months	2–3 months
average number in litter	7–8	6–8
length of life	2–3 years	18 months approx.

Rabbits have been greatly favoured as suitable pets for children for a long, long time, and it is therefore not surprising that these cuddly creatures are often seen in schools. Rabbits have been kept as pets in this country since the sixteenth century, and even today they continue to become increasingly popular.

The New Zealand White is possibly the best type for schools to keep, being easily obtained and very large so that children can move them about and generally handle them with complete ease—plenty of rabbit to get hold of. I am constantly surprised to find a large number of people who still think that the correct method of lifting a rabbit is by its ears, but this is far from true and is believed almost undoubtedly because of the way conjurors pull rabbits out of hats. Lift the rabbit gently by placing one hand under its chest and the other around its hindquarters—very gingerly at first, as a rabbit unused to handling tends to lash out with its hind feet and can inflict quite severe scratches on a naked arm. If it does kick, stroke it gently from head to tail and talk to it to try to gain its confidence, and then begin again. Rabbits are generally very placid creatures who like to be made a fuss of from time to time, but they do also like to be left alone to nibble titbits and watch whatever is going on around them.

The ideal place to keep a rabbit is out-of-doors (in many schools a practical impossibility) where it can enjoy the fresh air. If you are unable to provide it with outdoor accommodation, it should not be placed too close to a radiator as it is a very hardy creature who can very soon get too warm. Its home should take the form of a hutch, which some of the older children might like to design and make, or alternatively you can buy at pet shops at £4.75 to £8 each. If you intend to keep your rabbit outside, remember that the hutch will require some sort of insulation against very cold weather and, of course, damp.

Being herbivorous, or plant-eating, the rabbit will eat almost any sort of greenstuff from grass to cabbage and carrots, but always remember to give it hay too—no, not for bedding, which should be straw, but for it to eat. Many people do not know that hay is a very important part of a rabbit's diet, helping it to digest its other food. Rabbit pellets are also available at pet shops.

There are about fifty different types of rabbit for you to choose from should the New Zealand White not be to your taste. There are short-coated ones and long-coated ones (many of which require regular grooming), big ones, little ones, in many different colours. There are Belgian Hares, Flemish Giants and Californians (all large breeds), as well as Dutch, Netherland Dwarfs, Polish, Lops and Rexes, to name but a few.

When you come to clean it out—or should I say when the children come to clean it out—they will probably notice that it habitually uses only one corner of its hutch as a toilet. This toilet corner should be cleaned out every day and wiped over with a mild solution of disinfectant and fresh sawdust should then be sprinkled over the floor of the hutch. The whole hutch needs to be cleaned out only once or twice a week.

CAVY OR GUINEA PIG (Cavia porcellus)

Cavies, probably better known by the more popular name of guinea pigs, are, like rabbits, quite frequently seen in schools. They are, however, not so hardy as rabbits and during the cold weather should be warmly housed well away from draughts.

It is said that cavies never sleep, but personally I do not believe this to be true, although certainly they do appear to sleep very little. They also eat rather more than rabbits, this being quite understandable as they are much more active and therefore use up more food in energy. They require exactly the same diet as rabbits, only a little more of it. Baby cavies are born precocious (well advanced), having their eyes open, fur on their bodies, and teeth from the moment of birth, and are able to run about within a few hours. The gestation period is very long, about 63 days, and the litter normally averages three or four, although I have known as many as ten. Unlike rabbits, the boar and sow (male and female) can normally be left together, and cavies can be bred quite successfully on the colony system (one male to several females), although you do occasionally get one murderous individual who goes around killing everyone's babies, so do watch out for this.

There are many colour varieties, but only three main types of cavy: smooth coated, Abyssinian (the fur arranged in a series of rosettes), and Peruvian (very long coats that need grooming).

Cavies originate from South America, and are thought to be descended from the Restless Cavy (*Cavia pamparum*) now believed to be extinct. They were first brought to Europe sometime during the sixteenth or seventeenth centuries as an item of food. Just how this animal received the name 'guinea pig' is something of a mystery; 'pig' possibly because its flesh tasted like pork, and 'guinea' may be a corruption of 'Guiana', the part of America from which it was most likely shipped to Europe?

CHINCHILLA (Chinchilla lanigera)

Most people who think of chinchillas regrettably think only of fur coats and stoles and not of the living creatures which have to be killed to provide for their whims and fancies. However, this strange little creature does make a very interesting and unusual classroom pet.

Found in the Andes in South America, high above the tree line, this unfortunate animal has always been hunted for its fur, because it has the softest fur of any mammal. Although I must confess that I personally have done no research on this, I understand that fifty hairs grow from each hair follicle, and are so soft that you cannot even feel them. Try this experiment: hold your finger about 15 cm above the chinchilla and close your eyes, then start to lower your finger slowly; it will stop when it touches the animal's skin and you will not have been able to feel the fur as your finger passed through it. We have found that even blind children cannot feel it because it is so fine. It is also so dense that it is said to be incapable of harbouring parasites—and certainly I have seen no fleas or lice on any chinchilla I have met.

This creature is nocturnal, and generally speaking nocturnal animals are unsuitable for schools; however, chinchillas do not seem to mind being disturbed during the daytime—but do not let the children over-handle them. In their natural environment they survive on a very sparse diet of mosses, lichens, mountain grass, etc., therefore the diet in captivity should not be too rich; a little cabbage, carrot, sunflower seed, etc., is acceptable, or you can buy special chinchilla pellets from main pet shops or large department stores with a pet department. In the wild they keep their dense fur in good condition by taking dust baths, so provide a tray of Fuller's Earth Powder occasionally for this purpose.

Because chinchillas are very difficult animals to trap in the wild, earlier this century, in 1920, an American called Chapman struck on the idea of breeding them in confinement, and scrambled up the Andes to capture some; but because of the low temperatures and rarified atmosphere they had been used to, bringing them down to sea-level presented a major problem. This had to be done very gradually, so he brought them down a couple of hundred feet at a time and made camp for a few weeks, then brought them a little lower and made camp again, and so on; it took him two years to bring them from top to bottom. He was then faced with the problem of how to get them across the equator into the United States, so they travelled in specially refrigerated compartments on ships. The whole process took him three years, and it is from these specimens brought into confinement in the year 1923 that all the chinchillas we see today stem. The wild chinchilla is coloured silvery grey but we have now produced various colour mutations, although all of them are based on grey.

If you wanted a particularly good fur strain, you would have to pay £50 or more for each animal, but for school use this, of course, is not necessary. The ordinary pet chinchilla should cost in the region of £2 to £5.

FERRET (Mustela putorius furo)

If you want a really individualistic and independent class pet then, in my opinion, you will have a job to find one more suitable than a ferret. I was very much in two minds as to whether or not to include this particular animal, because it has a certain reputation for being unreliable in temperament and nipping from time to time; however, I have known many ferrets and their very close relations, and have not met one that, after gentle and considerate handling, regular feeding, a nice clean cage and a warm bed, has not become very tame and friendly.

The ferret is descended from the wild polecat, found in most parts of Europe, and appears in two colour varieties, the albino and the 'fitch', this latter being very much like a slightly lighter version of the wild ancestor. The ferret is very much a carnivorous animal and, left to its own devices, would feed on mice, rats, rabbits and insects, and might occasionally take a few chickens, together with a little fruit and vegetable matter from time to time. They normally grow to about 55 cm from nose to tail, a mere three or four centimetres shorter than the wild polecat. In some parts of the country they are used to hunt rats and rabbits, usually being muzzled in the latter case as they are by nature such extremely savage hunters that they are likely to slaughter several and remain below ground tearing them to pieces instead of driving others out into the master's waiting net.

Females, and to some degree males, do tend to smell once they become sexually mature, and if you do not intend to breed them it is often far better to have them 'doctored'. Like dogs, they are very susceptible to distemper and hard pad, and should therefore be inoculated against these diseases.

There are no problems with housing; a rabbit hutch is the best sort of home that you can provide, with a sawdust-covered floor, straw bed, and plenty of clean fresh water for them to drink. To minimize any smell, they should be cleaned out every day. Food should consist of whole organisms if possible, such as dead rats, mice, birds, etc., together with insects (mealworms) and a little fresh fruit and vegetables. If you are unable to feed whole organisms use fresh meat with mineral additives plus a little roughage in the shape of fur or feather. If properly looked after your ferret should live for seven or eight years.

DOMESTIC GOAT (Capra hircus)

Some schools are fortunate in being able to provide facilities for keeping animals somewhat larger than any of those previously mentioned. If you are able, and would like an animal children can see being really productive, may I suggest a goat?

Originating in Africa, the domestic goat is far from the hardiest of animals and will require to be kept well away from frosts and draughts on cold winter nights. There is certainly plenty of choice when it comes to selecting a specimen for the school, but here are the main breeds. Saanen (horned and hornless): this goat gives the most, but the poorest quality, milk and is invariably white; British Alpine: black with white facial markings (horned); Toggenburg: brown with white facial markings (horned); Anglo-Nubian: this type, as opposed to the Saanen, gives the least, but the highest quality, milk; it is the most delicate and the most attractive, with long, banana-like ears, and comes in a variety of colours, most often some shade of brown with a darker dorsal stripe.

If you want a goat merely as a pet, a billy-goat is the cheapest to buy and if he is castrated will not smell: if he is not castrated, however, I think that you will find the smell, especially during the breeding season, somewhat prohibitive to say the least. Incidentally, and I think I had better mention this, if you want a goat for milking then you will, of course, require a nanny; you may think this point quite elementary, but it has been known for a person who wanted a milking goat to purchase a billy and then wonder why nothing happened! Barren nannies, by the way, also make very good pets. Most goats have a life span of approximately fifteen years.

If you keep a billy-goat or barren nanny, feed them on herbage during the summer and hay in winter, with only a handful of concentrated food such as dairy nuts or, in certain areas, dairy mixture, or you can make up your own mixture consisting of bran, flaked maize, crushed oats and a little linseed cake (this is good for the coat). Beet pulp can also be included, but always soak it before feeding, as it is dried and you are liable to make the goat blow up if she drinks after consuming a large meal of unsoaked pulp. If you intend to keep a nanny for milk the same basic diet is required with a great deal more concentrated food, as your milk yield will be in direct proportion to the amount of concentrates consumed. Goats are not really grazers, although they do like to nibble at a reasonable amount of grass; they are fundamentally browsers, feeding on bramble, oak, willow and elm to name but a few of the trees and shrubs they enjoy. They also require plenty of fresh, clean water.

If you do not have any really high fences it is best to tether the animal for browsing, or you will probably get plenty of complaints from your neighbours whose roses, etc., disappear—goats are very good climbers and jumpers. Always tether on level ground or at the *bottom* of a slope, for, if you tether at the top, it is possible that the goat may lose her footing and roll downhill, hanging herself in her own collar and chain. Goats are very clean animals, and often very finicky, sometimes refusing to drink from water another animal has drunk from, and the drinking and food buckets must be kept scrupulously clean.

Because of their intolerance of cold and damp, it is best to bring the goat in at night, and a large garden shed or outbuilding is the best type of accommodation for her.

The mating season is between September and March, and if you want milk your nanny goat will, of course, have to be mated. If you do breed and decide to keep the youngsters, do not keep the sexes together after they are about two months old, as they are precocious animals and you might get inbreeding, which is undesirable, and anyway you would not want them to breed at such an early age. The gestation period is about five months.

Goats are very intelligent animals, and tough enough to stand up to a little rough treatment from inexperienced children. They will also stand up for themselves, giving the rough handlers a gentle butt, to show that they want to be treated with a little respect.

If the goat seems a little off-colour, cold tea is a very good general pick-me-up. The worst illness that your goat can get is entrotoxaemia, which is a killer as far as the goat is concerned, and a very swift killer at that. This is in no way contagious to man or any other animal, but if your goat looks 'droopy' or the droppings are loose, send for the vet *immediately,* because if it is entrotoxaemia every minute counts.

HOLIDAYS AND HEALTH RISKS

What to do with the school or class pet during holiday periods is one of the biggest problems that many teachers have to face. Indeed, I have known many schools that have refused to keep pets for this very reason.

Weekends during term time present little trouble, because many parents and pupils are only too pleased to provide temporary accommodation for a small animal, and, of course, many animals are quite happy left in a classroom over a weekend provided that they have been properly cleaned out and given enough food and water to last until Monday morning. This ideally should be made the last job on Friday afternoon. Before allowing any child to take an animal home, always check first with the parents to make sure that this is in order. Little Johnny will often tell you that mother has said 'yes', knowing full well that if he did indeed ask mother she would most certainly say 'no'. If mother really has said 'yes', then make out a list of the animal's requirements and give the child a quantity of food and bedding to take away with him.

If at all possible, do send the animal away for the weekend; if this is *not* possible, always make sure that the classroom it is in is kept locked, but make sure too that there is a key available in case of emergency. The reason that I suggest taking the animal away from school if possible is because most primary schools do not have a resident caretaker, and I know of one ghastly incident that befell one such institution in London not very long ago. Sadistic vandals broke into the school during the weekend and, not content with smashing and stealing equipment, they removed a class pet hamster from its cage and drowned it in a bowl of water in the middle of the classroom. The poor creature was found the following Monday morning by several children returning to their classroom before the teacher arrived, and they were, of course, horrified and heart-broken. Fortunately incidents like this are not too common.

Half-term and the long summer holidays need not be too much of a problem if you have a nearby secondary school. These schools usually keep a large quantity of animal and plant material that has to be constantly tended over the school holidays, usually a job for the laboratory technician who goes in either every day or every other day. Contact the head of the biology department and ask if he would mind your pet staying at his school during the holiday period; I am sure that he would be able to help, if only by suggesting alternate accommodation. However, do not leave making such arrangements until the last few days of term, but get in touch with him at least a fortnight in advance. Alternatively you could contact colleagues in neighbouring schools and form your own animal 'pool' by taking all your animals along to one school and making out a rota for their care—you could take it in turns. This method could have its complications if a number of teachers are going away at the same time so, if possible, contact the secondary school first.

Personally I do not think it a good idea for pupils to take animals home over long holidays unless you have a pupil who is particularly animal-minded and to whom you would feel quite safe in entrusting its care. Generally speaking, although children tend and care for an animal in class and perhaps over a weekend they will often become rather bored if they have to do this for a long period of time, especially if it is taking up some of their precious holiday. At first they forget to clean it out, then they forget to feed it and so on, and the animal suffers enormously because of this, so do please be careful.

When you purchase an animal for the class, try to avoid the ordinary 'pet shop'. While many of these establishments are excellent, there are a proportion who are interested only in their profit and will sell you even an animal on its last legs if they can. Local Education Authorities keep a list of biological suppliers and some even have their own distribution schemes; also buying by this method minimizes the risk of an animal being in poor health.

To be perfectly honest, there is far more likelihood of human beings transmitting disease to other animals than vice versa. However, there are two diseases that you should be made aware of; these are salmonellosis and psittacosis, more properly called ornithosis.

Salmonellosis is a bacterial disease found mainly in mammals, tortoises and a few birds, and causes enteritis (diarrhoea) in both other animals and man.

Psittacosis (Ornithosis) is a virus disease found in birds, although it is mainly restricted to members of the parrot family, fowls and pigeons. This disease causes a respiratory infection and can be very serious. According to one authority a very high proportion of the pigeons in Trafalgar Square are carriers of this disease although they show no apparent signs of suffering from it.

In my experience, many people seem to have an almost morbid fear of catching diseases from other animals. In this context it is as well to remember that *all* animals can contract more ailments from their own species than from any other—in other words, your pupils are far more likely to be infected by each other than by any pet you may choose to keep in your classroom.

If any animal shows signs of serious illness, do not attempt to dose it with drugs or preparations used in human medicine as these can prove fatal. Take the animal along to the PDSA clinic or a private veterinary surgeon and remember to keep it as warm and quiet as possible.

WAIFS AND STRAYS

Be they parents or teachers, every adult who is involved with children has, I suppose, from time to time been faced with the problem of how to cope with some strange denizen of the wild brought to them for succour. And let's face it, if we are on the one hand preaching conservation and preservation to our children, it is no good trying to shrug off such pleas for help on the other. It seems to us, then, that a few words on the care of those waifs and strays most likely to come your way would not be out of place here.

Broadly speaking, the casualties of the wild fall into two main categories, the orphans and the injured. When faced with the former it is as well to remember that, if you are to fulfil the role of rescuer fully and conscientiously, you are taking on a job for the rest of the animal's life, for it is seldom advisable to release hand-reared youngsters even when they have become more or less independent. I say this for two main reasons; in the first place, *you* cannot teach them as their parents would how to find their own food, avoid danger, cope with adverse weather conditions, etc., etc., and in the second place you *can* teach them the most stupid thing possible, that is to trust human beings. A fox cub, newly released, unable to find food easily, and consequently beginning to suffer the pangs of hunger, dashes hopefully up to the first man he sees—and receives not the hunk of meat he is hoping for but a charge of shot. Another problem in this field is to know when rescue is really required; warnings not to 'rescue' lost-looking baby birds have been widely publicized, because in fact many such apparent waifs are really only temporarily abandoned, probably while their parents search for food for them, and quite obviously the mother bird can bring up her young better than even the most well-intentioned human substitute. Do impress upon your pupils, then, the need to be really sure that help *is* required before butting in and carrying off some baby animal to be painstakingly reared in the home or classroom. In the case of birds, I think it would be safe to assume that fledglings really are lost if no parent has put in an appearance an hour after their plight was first noticed, and of course unfledged nestlings are always in need of help if the nest is pulled out by cats or other vandals.

Young birds are probably the easiest waifs to rear, being generally more than willing to accept food from any hand and also fairly catholic in their tastes. The types you are likely to meet fall into three basic types: seed-eaters, soft-bills (fruit and worm and insect eaters), and birds of prey. In spite of the fact that milk is strictly a mammalian food, I have known many nestlings to be successfully reared on a diet composed mainly of brown bread and milk, and certainly either seed-eaters such as finches, or soft-bills such as blackbirds and starlings can safely be started off on this food, with live food (earthworms, maggots, mealworms) interspersed and gradually making up the bulk of the intake. The main point to remember when rearing any young bird is that they require very frequent small meals throughout the hours of daylight, in order to cater for the requirements of their rather rapid metabolism. Also, of course, if nestlings are 'weaned' on bread and milk, their natural adult food, be it worms or fruit or seed, must be introduced into the diet gradually, so that their infant digestive systems can adapt without undue upset. Suitable diets for young birds can be catalogued as follows:

Seed-eaters: brown bread and milk; biscuit crumbs; porridge (as they begin to peck up food themselves); various bird seeds such as canary, millet, rape etc.

Pigeons: in a natural state, pigeons feed their young on partially digested food from their own crops, a substance known as 'pigeon milk'. It is, of course, necessary to emulate this when hand-rearing orphaned squabs, and this can best be done if you feed them upon chewed up biscuit, as your own saliva will naturally start to digest the food and thus provide a very fair imitation of the real thing. As the squabs normally feed by thrusting their beaks down the parents' throats, they will take food most readily if you are prepared to let them help themselves from your own mouth, but I must admit that this is a most nauseating proceeding which I would not recommend to my worst enemy! You may be able to get the bird feeding by letting it take chewed biscuit from your hand through your fingers, thus emulating the parent's beak, and then gradually lead him on to help himself from a dish.

Soft-bills: brown bread and milk; finely chopped apple, banana, melon, grape according to taste; soaked currants and raisins; mealworms (not too many of these as they are very rich); maggots (with the tough outer skin perforated as otherwise the digestive juices of very young birds seem unable to cope); earthworms, chopped up unless very small; beetles, spiders, and other small items of invertebrate life. Gradually the day will come when they can attack half an apple or orange or 'catch' whole large earthworms out of their dish for themselves and then your troubles will be at an end.

Corvidae (members of the crow family): these can be fed basically like the soft-bills, except that you should remember that they are very susceptible to a fatal ailment called gapes (a parasitic infestation of the trachea) which can be transmitted from earthworms, so this particular item of food should never be offered. As they grow older they will enjoy soaked dog biscuit, small chunks of raw meat, and dead mice, rats, or smaller birds, which they love to pull at and devour.

Birds of prey: from the time of hatching, birds of prey will eat only their natural food, that is dead animals from large insects up to chunks of chicken and rabbit. Roughage in the form of fur or feather is absolutely essential, so if you are unable to obtain whole unskinned animals you may have to resort, as I often have, to plucking feathers from the cushions with which to adorn the raw meat. Allowing for individual variation, birds of prey seem to be rather less confiding than most species, so you may well have to force-feed your protégé to start with, by holding the beak open while you poke scraps of food down its throat; this is most easily achieved by two people working together, one operating the beak while the other inserts the meal.

On the whole, young mammals are far more difficult to rear than birds; for one thing, they will often flatly refuse to take food at all, in which case there is precious little you can do about it, unless you happen to have on hand some friendly disposed cat or bitch, already lactating, who is willing to take on foster children. I have known a bitch suckle a young goat and a domestic cat bring up a dingo pup, but such accommodating behaviour is unfortunately fairly rare. However, there is nothing like 'having a go', so if you are presented with a suckling squirrel or fox or badger or whatever, it is at least worth trying your hand at rearing it. Your first problem will be to find a feeding bottle of the right dimensions—for small individuals I have found a doll's bottle quite satisfactory, or you may be able to obtain one of those especially designed for premature babies from a chemist. Failing either of these, an old-fashioned fountain pen filler or eye-dropper will often answer your purpose. The next thing is to work out a suitable composition for the 'milk'. Many people make the mistake of thinking that diluted cow's milk is the answer to everything, on the grounds that a cow, as a large animal, must necessarily have stronger milk than diminutive creatures such as rabbits and squirrels. Nothing could be further from the truth, for the cow has in fact about the weakest milk there is, so you are much more likely to have to strengthen it than dilute it, as the following table of milk analyses will show:

I have found a good recipe for hand-rearing puppies to be made up of full-cream baby food and slippery elm food, mixed in the pro-

	water	*fat*	*protein*	*sugar*	*salts*
Cow	87·1	3·7	3·75	4·85	0·6
Goat	84·0	6·25	4·0	4·75	1·0
Squirrel	72·4	12·6	9·2	3·4	1·4
Fox	81·93	6·30	6·25	4·56	0·96
Hedgehog	79·4	10·1	7·2	2·0	2·3
Rabbit	69·5	10·45	15·54	1·95	2·56
Dog	75·5	12·0	8·0	3·1	1·2

portion of two parts to one, and one teaspoonful of the mixed powder added to the cream off a pint of milk together with a teaspoonful of honey. If we take this as the basis for our enriched milk and, by reference to the comparative table above, boost the sugar content by the addition of extra honey or the protein content by a little beaten egg, we can at least approach the constitution of the milk required for any of the waifs we are likely to have brought into our care. It will be seen from the table that goat's milk is considerably richer than cow's, so if we already have a goat as a school pet, our problems may be minimized. The addition of a couple of drops of Abidec (a multi-vitamin preparation obtainable from the chemist) will bump up the vitamin content—always supposing that it does not put the infants off their food altogether, as it tastes singularly foul!—and I have found that a little lime water (an old-fashioned, but nevertheless extremely efficacious source of calcium, also obtainable from the chemist) together with the cod-liver oil which is essential for its absorption, will ensure strong bone formation and avoid all danger of rickets.

Should you be presented with an infant of a less commonly 'rescued' species, you can often approximate the correct constitution for the milk by choosing that given for the closest relative; a weasel or a badger, for instance, being carnivorous (a meat-eater), is more likely to thrive on bitch's or fox's milk than squirrel's, although the case of the badger is in fact rather specialized owing to its need for an exceptionally high iron content which necessitates the addition of neo-ferrum, or some similar product, to the basic recipe if the foundling is not to die from enteric disorders.

When it comes to weaning, it is wise to be guided by the youngsters themselves, for they have the best idea of when they are ready for solid food, and there is great individual variation on this point. Offer a little of the appropriate solid food shortly after the eyes are opened, and continue to do so every few days until interest is shown. Remember, too, to offer a variety of foods, as other animals have as many different tastes as we do, and one rabbit may accept turnip while declining carrot or vice versa. In the case of young rabbits, incidentally, it is advisable to avoid any item not completely free of soil, as this carries the dreaded coccidiosis, which is absolutely fatal to young stock. As with birds, little and often is best in the early stages, although on the whole young animals are more considerate than human babies and 'round the clock' feeding can soon be discontinued!

Elementary first aid is often necessary for your own pets as well as injured wild animals which may be brought in, so it is as well to acquire at least some knowledge of the subject before you embark upon keeping even a solitary hamster in your classroom. Eye troubles seem to be among the most common disorders in animals kept in confinement, often being caused either by draughty quarters or some foreign body, such as a scrap of straw or hay, penetrating the eye. In either case, bathing with a solution of one teaspoonful of common salt to the pint of boiled water is often effective, or a little (only a little, as it is mildly poisonous) Golden Eye Ointment squeezed actually into the eye may be of use if the trouble persists. Enteritis is another common ailment; this may be diagnosed by copious faeces, blood-streaked in really serious cases, in mammals, or by green faeces and an accumulation of same round the vent in birds. In the former case, terramycene administered orally is often effective, and in the latter sulpha-mezzathine added to the drinking water can work wonders, as, indeed, it can with most avian ailments. Egg-binding sometimes occurs in female birds, and can be diagnosed by a general air of lassitude and malaise. Anointing the vent with warm olive oil, or holding it over a gently steaming kettle will sometimes release the egg which is causing the trouble, but extreme cases should be referred to a veterinary surgeon as a minor operation may be required.

Probably the most common injury with which you are likely to be faced in a wild animal is a broken limb. Birds' wings can seldom be set so perfectly that they are able to regain the power of flight, but the bones may be induced to knit if the wing is strapped into position against the body and held thus for a few weeks. A very satisfactory splint for a broken leg in a small animal can be made from an ordinary pencil, cut in half longitudinally and with the lead removed, the limb fitting neatly into the resultant grooves. All bandaging aimed at holding broken limbs in position should be with sticking plaster, as non-adhesive bandage is apt to slip out of place, especially on the tapering legs of birds. Really badly injured limbs in small birds can often be amputated very easily with one snip of a pair of sharp scissors, but injuries necessitating stitching or real surgery should, of course, have the attention of a veterinarian. It is also as well to remember that birds' blood does not clot as readily as mammals', so heavy bleeding can often be stopped only by the application of a tourniquet, although a copious coating of Newskin may sometimes check haemorrhage.

Especially in very small animals, shock is often the deciding factor between life and death, so do remember to keep any patient warm, quiet, and if possible in the dark, when the worst effects may possibly be allayed.

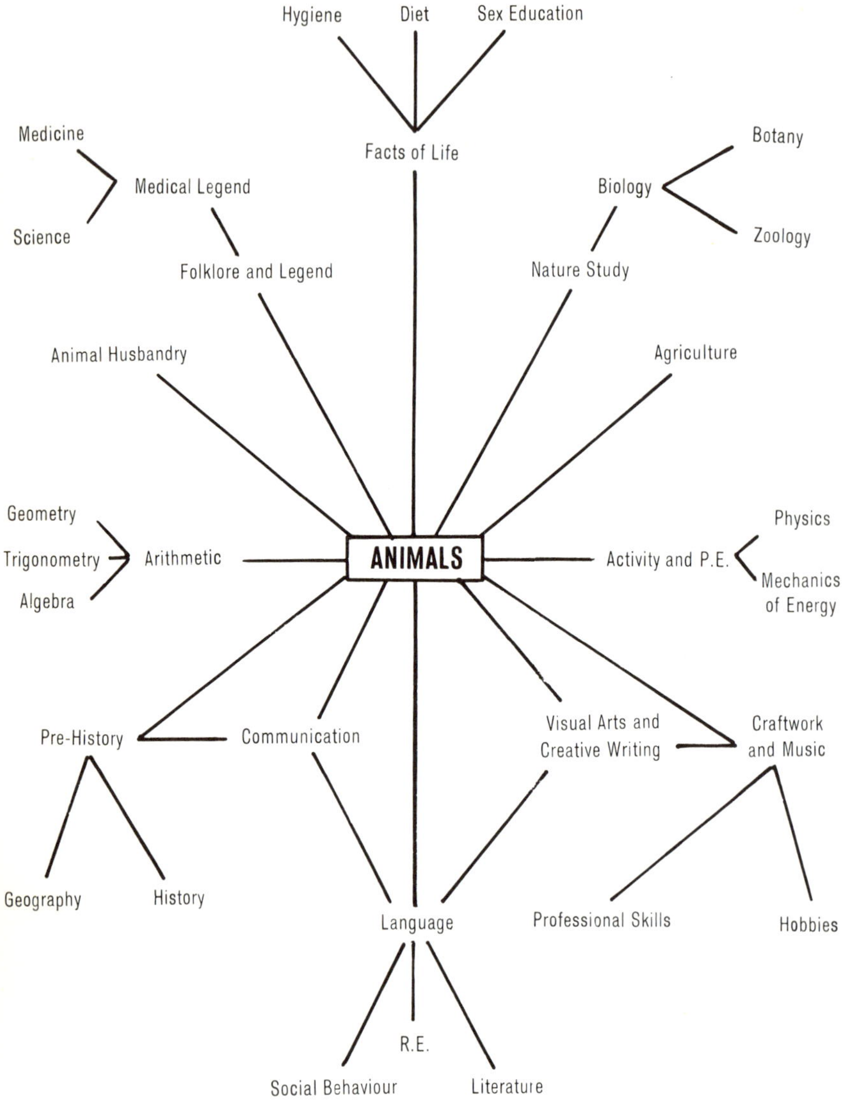
Hygiene
Diet
Sex Education
Facts of Life
Medicine
Medical Legend
Science
Folklore and Legend
Biology
Botany
Zoology
Nature Study
Animal Husbandry
Agriculture
Geometry
Trigonometry
Arithmetic
Algebra
ANIMALS
Activity and P.E.
Physics
Mechanics of Energy
Pre-History
Communication
Visual Arts and Creative Writing
Craftwork and Music
Geography
History
Language
Professional Skills
Hobbies
R.E.
Social Behaviour
Literature

The use of animals in class

In the foregoing pages we have recommended a good many different species of animal which we have found suitable to be kept as classroom pets. The majority of primary schools do already, in fact, keep at least some of them, and have presumably paid considerable attention to their correct care and maintenance. But in most cases, in my experience, the matter stops there. A solitary goldfish swimming miserably round and round his bowl in one class, a somnolent hamster skulking in his bedroom in another, possibly a few stick insects moribund on their twigs of withered privet in a third—this is what one tends to find in schools all over the country. One gets the impression that, while the initial idea of living animals in school had immediate appeal, once they were

A somnolent hamster skulking in his bedroom

installed nobody had any ideas about how to use them, and interest gradually waned until even looking after them became a tiresome chore performed less and less willingly and less and less frequently.

What a potential goldmine the innocent teacher is sitting on! I have tried in the following chapters to suggest just a few ways in which animals can be used as 'pegs' upon which to hang all manner of learning, in the hope of whetting your appetite to pursue further ramifications on your own initiative. The possibilities are endless, the animal kingdom is vast—'ask now the beasts and they shall teach thee!'

Let me offer two words of warning, though. Do not be tempted to base your entire teaching programme upon the animals at your disposal; this would almost certainly be possible, but one can have too much of a good thing, even animals, and you do not want your pupils to become as bored with your snake, ferret, budgerigar or whatever as they are already likely to be with their six-times table. Also bear in mind that, while animals can be *used* to excellent advantage in schools, they should not be *abused*. Other animals are just as much individuals as the rest of us; they have their likes and dislikes, their 'off-days', their little fads and fancies just as we do, so, as well as attending always to their physical requirements, do take care that they are not turned into nervous wrecks by a surfeit of petting and handling, no matter how well-meaning it may be. If your Common Boa makes it clear by hissing that he does not feel like being picked up and cuddled today, take his word for it and let him alone; if your gerbils refuse to come out of the nest and metaphorically hang a 'do not disturb' notice on their bedroom door, tell the children to obey their pets' obvious wishes, which may anyhow mean that babies have been born, a time when any disturbance is likely to drive their mother to infanticide. Provided all the children understand fully that the animals are not only dependent upon them for their daily care, but also have their own rights as individuals, everything should go swimmingly and the educational use of pets become a pleasure not only to both you and your pupils but to the animals as well.

You will have noticed, too, that, with very few minor exceptions, we have advocated no British animals as suitable classroom pets. This was deliberate. The World Wildlife Fund and other similar organizations are constantly bringing to the notice of the public the urgent need for conservation of wild life all over the globe, but this overcrowded little island of ours is almost certainly in greater need of such a policy than anywhere else. Here our ever-growing hoards of *Homo sapiens* are rapidly wiping out all our wild fauna, trapping them, spraying them with poisons, or more simply destroying their habitat. It is for this reason that, apart from a few small items of pond life, we have not recommended collecting your own specimens; even the all too common practice of hatching frog spawn in the classroom is to be deplored *unless* the youngsters are returned to the wild at about the time when they are ready to come out onto dry land. Very, very few of the myriad jam-jar hatchlings which are produced annually in both home and school live to reach maturity, largely because of feeding problems, and this is one of the reasons why the frog is now classed as a rapidly vanishing race in this country. If, of course, you are prepared to collect and hatch spawn, and then take the trouble to rehabilitate the tiny frogs in an area where they are relatively safe from loss of habitat, you will not only be actively helping to preserve the species but will also have the benefit of an easily observed development cycle for the instruction of your class.

'THE FACTS OF LIFE'

The vast majority of the problems and difficulties which beset twentieth century Man on all sides might well be mitigated if not dispersed altogether could he but remember what he is—a member of the animal kingdom. Certainly this applies superlatively to all the web of complications we have laboriously woven about what are euphemistically known as 'the facts of life', and it is undoubtedly by seeking the help of other animals that these can most effectively be smoothed away. The introduction of sex education into schools, while incurring the wrath of some, is undoubtedly a step in the right direction. But how best to accomplish this in the very young? For it is the very young who ask most questions and, so the psychologists tell us, these should be answered clearly and truthfully at the time of asking, not five or ten years later, when the unwilling recipient of the information has probably given up his quest for knowledge and lost interest altogether.

I suppose my own children were lucky in this respect, for they grew up among animals, and thus all knew not only where babies came from but how they got there long before they went to school. I remember one, at less than three years old, deducing just how he himself came into the world when he saw a litter of kittens born; another at similar age asked indignantly whether I wanted the mare to have *another* foal—'Well, the stallion is mating her!'; while yet a third set me quite a poser by demanding to know why the father of an unborn child did not know its sex since 'he put it there'. How, for heaven's sake, does one explain sex-determination of the developing foetus to a five year old? Later on, of course, when they could really *understand* such matters, we delved into the mysteries of chromosomes, menstruation, pregnancy, and even venereal disease as and when they cropped up—but this is jumping ahead. To me, the most important 'fact of life' which can be inculcated even into nursery classes is the wonder of the regeneration of life; that, far from being rather dirty and quite unmentionable, it is a bodily function little short of a miracle, and above all *natural*—as natural as breathing and eating and walking—yes, and going to the lavatory!

As we ourselves are mammals, obviously mammals are the animals most suitable for use in this field. Live-bearing fish (swordtails, guppies, mollies) might provide a useful start, as they can be clearly seen and studied, and it is always quite apparent when the females are gravid, as the developing foeti form a distinct dark 'blob' in the distended abdomen. Unfortunately, it is very seldom that one is lucky enough to see the mother actually give birth—in fact, although I must have bred hundreds, I have never managed to be present at the moment of parturition. Also, even though you have dutifully isolated the gravid female (most easily accomplished by floating a half-filled jam jar in the aquarium to serve as a maternity ward), it is more than likely that she herself, apparently feeling peckish after her labours, will turn round and devour most of her new-born young before you realize what is happening. This is not only disappointing, if not heart-breaking for the children, but can also give rise to a number of wrong ideas, such as that the mother-love on which they themselves depend so much is only an empty shadow. They are still too young to understand that maternal care is for the most part a comparatively late development in the animal kingdom, or to appreciate that, in a natural state, the young fish would immediately swim among the cover of plants and rocks and so hope to escape the attentions of their parents and other relatives.

I suppose that the common frog is used more than any other creature as an illustration of the development of new life from almost microscopic beginnings. Whereas the shell of a bird's egg is opaque, so that the foetus growing inside it during incubation can be seen only if it be killed and the egg smashed, the tiny black speck of a frog's egg can be watched daily as it gradually enlarges and changes shape until the tadpole itself is ready to wriggle free from its jelly-like protective covering. But there are snags about this species as a classroom illustration too. Every year thousands if not millions of tadpoles are successfully hatched out in schools all over the country—but how many of them reach maturity? Generally speaking, the eggs may be satisfactorily observed turning into tadpoles, and the tadpoles into minute frogs—but here the trouble really starts. While the spawn requires no specialized treatment other than fairly clean water and an even temperature, and the tadpoles will readily devour a wealth of foodstuffs ranging from algae to scraps of raw meat, young frogs are extremely difficult guests to cater for, as they will eat only the tiniest of living insects, so, unless you are prepared to indulge in the fairly complicated task of breeding your own supply of Drosophila (fruit flies), you would be well advised to return your baby frogs to their natural habitat as soon as they are ready to venture forth upon dry land. There is also the very serious question of conservation; the common frog, partly owing to loss of habitat, but mainly, I fear, owing to the annual depredations of fishers after frog-spawn, is now a rapidly vanishing species in our own country, and it is for this reason that I earnestly entreat all teachers to resist the temptation of decorating their classrooms for a few short weeks with attractive tanks of spawn and tadpoles, even though they had intended to exploit its educational potential to the full.

No, choose a mammal if you intend to develop the theme of sex education. For this purpose I think that the domestic cavy (*Cavia porcellus*) or guinea pig is almost certainly the best illustration for a number of reasons. The adult pair can normally live together throughout, and are large and usually phlegmatic enough to withstand handling, weighing, and measuring during pregnancy, which is satisfactorily apparent. The young are born precocious, so can be watched closely before they are dry and even examined within a few hours of birth, unlike baby rabbits, rats or mice which should not even be seen for several days at least. Growth rate in the cavy can be observed from the first day, the young can be seen suckling—in fact the whole sequence of reproduction in a typical mammal from mating to weaning can be graphically unfolded before the wondering eyes of your class.

Whichever living animal you choose to illustrate your teachings on this particular subject, you can, by your own approach, inculcate not only basic biological knowledge but a positive and healthy attitude towards life itself, by quashing the embarrassments and inhibitions which have done so much damage in the past. No girl who has actually *seen* the miracle of birth should envisage childbirth as a terrifying experience nor sex as a disgusting interlude which must be 'put up with' for the benefit of the male. I have known women of my mother's generation literally do themselves physical harm by refusing to admit that they needed to visit the lavatory when in mixed company, but now the children can be shown that all animals have lavatory habits in the interests of hygiene (notice how your cavies tend to soil only one part of their hutch, well away from their beds!) and not because this natural function is something to be ashamed of. 'Ask now the beasts and they shall teach thee'—in this context I feel that this quotation might be re-written as 'Buy now your class pets, and let them teach for you!' In other words, there are more basic 'facts of life' to be inculcated into the growing child than those of sex alone,

Buy now your class pets, and let them teach for you

and although we tend to think that only the human species is aware of the importance of hygiene, we have only to look really closely at almost any animal to realize that this is far from the case. Everybody has exclaimed appreciatively at the attractive sight of a cat washing herself, but how many of us, I wonder, have noticed that rats and mice, cavies and hamsters pay equal attention to personal cleanliness. Quite apart from the obvious risk of disease, regular washing is essential to the perfect functioning of most external parts of the body; our skins cannot 'breathe' or sweat if the pores are caked with dirt, we cannot breathe properly if our nostrils are blocked with mucus, and our hearing is impaired if our ears are allowed to become coated with wax and grime. Similarly the sensitive ears and noses and whiskers of a mouse or a cat cannot function adequately unless their owners are constantly cleaning them—and in their case imperfect functioning of any single sense may mean the difference literally between life and death, not merely between being a good or bad conversationalist. Let the children watch the mice polishing their whiskers and the rabbits 'combing' their ears, and lead them on to draw the obvious inferences. But let them notice too that the whole body comes in for a regular cleansing—indeed I am sure that most animals pay more attention to bodily cleanliness than do a great many human beings. A dirty coat will tend to absorb moisture instead of providing a waterproof protection against it, so here again it is vital for birds and mammals at least to adhere rigidly to

Most animals spend a lot of time washing themselves

the fundamental rules of hygiene. Your animals will also help you differentiate between types and degrees of dirt; fresh, unadulterated soil is not dangerous dirt in the same way that excreta is, other than in the comparatively rare instances where it may harbour tetanus or anthrax, either of which can be contracted only through a cut or broken skin. In fact dry earth can supply a cleansing agent in much the same way as a dry shampoo, as can be witnessed by the chinchillas who roll and squirm happily in their Fuller's Earth Powder. The children can learn from this that if their hands become contaminated with some noxious filth when they are far from home on a country walk, they can to some extent cleanse themselves by 'sandpapering' the impurity away with a handful of dry earth.

And what about diet? Personally I feel that the rudiments of dietetics cannot be inculcated too early, even if it be simplified to the four most basic essentials, protein (body building), carbohydrate (energy), vitamins and minerals. The body of every living animal is built up of and functions by means of these ingredients, and good health cannot be maintained unless all are present. Different types of animals have digestive systems adapted to satisfy these needs from different sources—a herbivore or plant-eating animal requires a greater length of gut to acquire its nutriment from herbage than does a carnivore (meat eater) or omnivore (general feeder)—and different types of teeth are required for different foods. As general feeders we ourselves possess a varied dentition, incisors ('nibblers' at the front of the mouth), canines (seizing and tearing) and molars (grinding). Let the children watch how the mice nibble rapidly, the ferret tears and gulps, and the goat grinds placidly away as she ruminates. Protein *can* be obtained from herbage provided the protective cellulose covering is sufficiently broken down (hence the goat's chewing and rechewing) and carbohydrates and minerals reach the carnivore if a whole carcass be devoured. Many pet owners who dotingly feed their meat eaters on best steak alone thereby sign their death warrants by depriving them of the carbohydrate and minerals to be found in bone, fat, and the contents of the gut. Even earthworms provide a more complete diet for the classroom terrapin than a diet entirely composed of lean meat and, of course, vitamin content is almost synonymous with freshness—snatch the opportunity to point out that a single raw apple has more nutritive value than a whole plateful of overcooked, sloppy cabbage. Mice and rats are probably the best teachers when a balanced diet is the question under discussion. Offer them small quantities of a large selection of foodstuffs—raw meat, grain, cheese, fresh vegetable matter—and note how they will sample a little of each rather than devour all of one commodity and ignore the rest. This, of course, is an experiment which will work only if the animals have been accustomed to a balanced diet, as those who have been previously deprived of any essential will tend to go mad over that one item to the exclusion of all others. It is also of value to point out that other animals do not run to obesity; with the exception of a few domesticated species, notably the dog, who have become contaminated by the human mentality, animals never have to go on a diet for the very simple reason that they eat only enough for their bodily needs—they do not, in fact, 'eat for eating's sake', which is the downfall of so many human beings. I am not suggesting that some high philosophical ideal is behind this frugality; the reason is simple and obvious. In the wild the process of feeding is either difficult (as for the predator who must stalk, and watch, and wait, and run to acquire his dinner) or dangerous (as for the natural prey who are in mortal danger all the time their attention is focused on the act of eating), so that it is small wonder that they stop just as soon as they have absorbed the barest sufficiency.

Possibly the most important 'fact of life' which can be inculcated via the medium of the classroom pet is a social one. Children are fundamentally selfish animals, paying little heed to the needs or, indeed, even the rights of others, an attitude which must be altered pretty radically if they are to take their place as useful citizens in twentieth century society. Although the ultimate responsibility for the well-being of the classroom pets must, of course, rest with the teacher, this should be delegated whenever possible and the necessary supervision kept as unobtrusive as may be. Stay behind for ten minutes after the children have gone home to check that they have indeed shut up the birds safely, watered the goat and given sufficient fresh food to the rabbits—but *let them do all these things by themselves* with no over-anxious adult breathing heavily down their necks. A child who can be, or believe himself to be, wholly responsible for another life, be it no more than that of a stick-insect, is well on the way to developing a sense of responsibility towards other members of his own species, an attribute which should stand him in good stead for the rest of his life.

One final lesson which may be taught through your pets is on the subject of poisonous plants. As we have already said, fresh food is always the best food for any animal, so during the summer let the children 'forage' for their pets' fodder whenever this is possible—a task which can be educational and help your pocket at the same time. Rabbits and cavies like not only grass, but many of the umbellifera which abound in our meadows and hedgerows, and here is a heaven sent opportunity to discourse upon the dangers of hemlock, with its ultra-tall growth and smooth stem. Vetches supply valuable protein, but beware of the laburnum pods which are so similar in appearance. Dandelions are appreciated, especially by tortoises, while growing buttercups are poisonous, although harmless in hay. Gerbils, rats and mice, and most of your birds appreciate fresh berries—hips, haws, rowan, etc.,—and in the search for these attention can be drawn to the identification of deadly nightshade and the eminently toxic yew. Chickweed and shepherd's purse, to be found growing as weeds even in the heart of the city, provide valuable greenstuff for budgerigars and small rodents, but the so attractive foxglove contains digitalis and must be avoided like the plague. Broadly speaking, most evergreen shrubs are poisonous and best avoided, although stick insects of different species thrive on a diet of privet, ivy and rhododendron, and goats will often take a small quantity of ivy for medicinal purposes. Laurel, however, should *always* be avoided, as its toxic content is so great that it can actually be used as the lethal agent in killing bottles used for the destruction of various insects required as display specimens.

ARITHMETIC

A friend of mine, who is a highly intelligent biology master in a secondary school, once told me that he could not begin to comprehend mathematics unless it was first related to animal life! And, incredible though it may seem, this is one of the easiest subjects to teach at all levels via the medium of living things. For simple addition and subtraction mice are probably the best subjects; a pair of mongrel-bred parents of different colours will almost certainly produce pretty varied litters—4 black, 1 white and 2 grey make 7

Weighing and measuring the class pets in a science lesson

in all; take away the grey youngsters and you have only 5 left, and so on. We started with 2 mice who had 7 babies, so now we have 9; John and Mary each want a pair to keep as pets at home—2×2 (4) will therefore be taken away, leaving us with—how many? Use your millipede for multiplication: there are eighty segments on her body, each carrying two pairs of legs, so how many legs has she in all? Two pairs $= 2 \times 2 = 4$, and $4 \times 80 = 320$. Many children are foxed by an attempt to do such seemingly huge sums in their heads, so seize the opportunity to reduce the arithmetic here to common-sense level, and point out that $4 \times 8 = 32$, which they all know, and 4×80 requires only the addition of a nought to that simple application of the four-times table. The ramifications are endless. Weights and measures experiments are probably best carried out with cavies or rabbits. Growth rates can

Children digging a trench for electrical cables for their out-door pet enclosure

be determined by regular weighing and measuring of young stock, as can the increase of girth and weight occasioned by pregnancy. How many square metres of lawn can a cavy 'mow' in one hour, I wonder? How far can a tortoise walk in five minutes, or how many ounces of cabbage will it consume in a day? Does a bird's egg weigh the same when it is laid as it does after a week's incubation, and if not, why not? Incredible what a large part mathematics play in our everyday pursuits of eating, moving, and breeding!

All this is fairly straightforward and obvious; so fundamental, in fact, that virtually any animal you may decide to keep can be used mathematically. The children can count the stick insects' eggs, and then see what percentage of them (a) hatch out at all, and (b) reach maturity; or reduce to fractions the variations in size between the adult and the baby, the male and the female, or even two different species of animal altogether. *Average* weights, heights, lengths may be ascertained if several animals of the same type are kept, and even the cash value of the creatures together with that of their food and progeny, may be utilized in the teaching of pounds and pence. Some of the amazing mathematical facts of nature can be used to arouse the children's interest in the first place; did *you* know, for instance, that certain primitive races use the size of a bee's cell as a unit of measurement, with such mathematical precision do these insects construct their honeycomb? Or that the Brush Turkey can gauge the temperature of her mounded nest to within two degrees Fahrenheit by using her naked neck as a thermometer, much as the traditional nurse-maid tests the baby's bathwater with her elbow?

If you start keeping animals from first principles, however, applied mathematics (yes, even at primary level!) can also be taught, together with the rudiments of architecture, estimating, design, and carpentry. Let us suppose we have decided to keep some ducks, and have the headmaster's approval to go ahead within certain financial bounds. What is the first thing we do? It is obviously no good simply scanning the advertisements for the animals of our choice; first we must learn all we can about the requirements of ducks (or canaries, or rabbits, or snakes, or what have you) before moving on to the first really practical step which is the design of the animals' future home. It is amazing how quickly and eagerly even young children will set to work to draw up their very first essays in architecture—and also how soon, with a little help and encouragement, they will come up with a practicable outline. Obviously this roof will need more support, or that door will not open unless the hinges are put on the other side; but problems can soon be ironed out and a final plan achieved which is not only pleasant to look at, but which will also (we hope!) *work*. Now—what materials do we need? Let the children work out in detail just how many centimetres of what type of wood will be required, how many metres of wire-netting, even how many screws, hinges, door-fastenings, and nails—a quick trip to the iron-monger's will tell you roughly how many nails there are to the kilo—and from this we can progress to accurate costing of the whole project.

Assuming we receive the go-ahead, the materials can now be ordered, and the children set to work to construct the actual 'residence' according to their own design. Personally I see no reason why this stage of the operation need necessarily be restricted to the boys in the class, as most women are better off for knowing how to handle a screw driver and drive in a nail straight! Sawing, hammering, screwing and snipping, the children will probably surprise you with the efficiency of their efforts. Mistakes will be made in plenty, but we all learn by our own mistakes after all, and many a future carpenter will be grateful for the lessons gleaned from his struggles with his first rabbit hutch.

Mistakes will be made in plenty

Different types of animals require different accommodation; an obvious platitude, this, but it is not, perhaps, so immediately obvious what different subjects may be taught in this field depending upon the initial choice of specimens. Ducks need a pond; in what proportions does one mix concrete, or how 'puddle' clay? Any animals requiring to be housed in vivaria will entail learning how to set glass into wood; it is even possible to construct a water-tight aquarium from these same materials! Burrowing animals will involve you in different problems from flying animals; species which gnaw, such as mice and hamsters, are more difficult to accommodate than those which do not. All in all, the mathematician and architect have unlimited scope in their dealings with the animal kingdom, and there is no reason why all their knowledge and its application should not be handed on to the pupils via the same medium.

ENVIRONMENTAL STUDIES AND SOCIAL SCIENCE

I must admit that I am still old fashioned enough to think of these subjects as geography and history, but at least I can admit truthfully that modern methods do more to bring them to life than did the drawing of geological maps and learning of endless lists of dates to which I was subjected at school. And in geography animals can, of course, be of enormous educational value, since they not only share our environment with all its hazards and benefits, but are themselves a part of it.

The animals included in this book as suitable subjects for classroom pets are not, of course, all the possibilities—I actually know of schools which keep such comparative rareties as caimens, steppe lemmings and edible dormice, for instance. Geographically speaking, however, even this 'short list' covers creatures found not only in our own country, but in India, Corsica, Madagascar, Java, Australia, South America, Africa, Thailand, Central Europe, Mexico, USA, Spain, Portugal, Greece, Italy, Morocco, China, Syria, Scandinavia and Japan!

Let us take just one type of animal from each section and see how it can be used as a 'peg' on which to hang a wealth of historical and geographical teaching. The leech is an excellent example. Assuming that the pupils have already learned a good deal about the habits and way of life of their classroom pets, they will be able to deduce a good deal about the country they came from—in this case the locality must be one well supplied with water in the shape of slow flowing streams or ponds, which in turn suggests a fairly flat terrain, far from the rushing torrents of the mountains, coupled with a moderately damp and, in the case of the Horse Leech, cool climate. From these beginnings it is a simple matter to diverge on to the type of vegetation one might expect to find in these conditions—for instance, sedges and marsh marigolds and lady's smock instead of heather, stonecrops and other plants which demand a more arid soil. Historically the value of the leech is obvious, for were not certain species used as an almost universal nostrum for all ills when the practice of medicine was in its infancy and 'bleeding' held to be the only panacea for 99 per cent of ailments? How leeches were replaced by the barber with his cup and knife and he in turn by more varied (and effective) forms of treatment might also form the theme of a treatise on medical history. Strange how all human learning can be inter-related, for here we can aptly digress into the realms of literature with the following lines from Shakespeare's *Macbeth* (Act IV, Scene 1):

> Fillet of a fenny snake,
> In the cauldron boil and bake;
> Eye of newt and toe of frog,
> Wool of bat and tongue of dog,
> Adder's fork and blind-worm's sting,
> Lizard's leg and howlet's wing,
> For a charm of powerful trouble,
> Like a hell-broth boil and bubble.

Macabre medicine, maybe, but none the less not far removed from man's first strange essays in this direction.

Let's turn now to fish—the humble goldfish for instance. No zoological detective work to be done here, for this is essentially a domestic animal, artificially bred to delight the eye of the pet-keeper. The common carp, from which it is descended, is a native of the chilly ponds and streams of Europe, where it has been esteemed as a gastronomical delicacy down the centuries. Before the dissolution of the monasteries in the reign of Henry VIII ponds well stocked with carp kept the monks happy in their weekly fasts throughout the middle ages and, although I have heard the flesh likened to cottonwool stuffed with needles, presumably those ancient priors and their dependants found it most palatable. Still classed as a 'coarse fish' by those who hunt this type of animal for sport, the carp is a remarkably tough species, which can be transported for long distances in no more liquid surroundings than damp grass or moss, which is perhaps why the Japanese hit upon it for some of their early experiments in the breeding of strange 'sports' from a wild progenitor. Having first, by selective breeding, produced the gold colouring in what should naturally be a silvery-green creature (a colour still apparent in very young specimens), this ingenious and artistic race went on to meddle with the very shape of the animal, until it now appears in a great many widely differing forms, both beautiful and grotesque. Veiltails, comets, black moors are only a few of the types now available in most pet shops, with their flowing caudal fins, or goggling telescopic eyes, and of course the scaleless shubunkin, mottled in gold and pink, is yet another ramification of the same involved family tree. Unfortunately many of these 'sports' are shortlived owing to an inbred tendency to swimming bladder disease which prevents them maintaining an even keel, but this does not appear to have in any way affected their popularity.

Among the amphibians the salamanders probably arouse the greatest social and environmental interest—although the poison distilled from certain South American frogs for the arrowtips of primitive tribesmen might well lead on to a good deal of research into the ancient civilization of this continent. Of the two species recommended for schools the axolotl can best be used as a basis for geographical explorations, while the Fire Salamander leads us into the realms of history, myth and legend. Amphibians, we already know, require damp conditions, so let us deduce why in the particular area where the axolotl is found the Tiger Salamander elects to live its entire life in the larval or tadpole form.

The fire salamander leads us into the realms of, for example, biology

Obviously local environmental conditions must in some way prohibit its development into an adult—and we do in fact discover that Mexico is an exceptionally arid country, poorly suited to sustain amphibian life. There must, of course, be ponds and streams, as otherwise the entire country would be no more than a desert, but let an amphibian once venture away from the water where it was born, and it will sign its own death warrant through desiccation. Thus, nature being nothing if not adaptable, the Tiger Salamander in Mexico has learned to stop short at the larval stage, when, as an axolotl, it reaches sexual maturity and is thus able to reproduce itself together with its highly specialized method of overcoming the problems of this particular environment. The Fire Salamander of Europe faces no such problem, for no-one could call the climate of this part of the world too dry! But how, I wonder, did it get its name? Apparently our forefathers were remarkably lacking in powers of observation, because they sincerely believed that these fairly typical amphibians were born in fire —the very element most violently at variance with their natural habitat! It is just possible that this strange myth sprang up because of the animal's improbable colour scheme, the bright yellow, or in some cases orange or red, spots and blotches being reminiscent of patches of fire, but it seems to me more likely that the salamander's true mode of life may, oddly enough, have been responsible. Like most amphibians this species is of a retiring nature and, being nocturnal, tends to hide away by day, under stones, under dead leaves, and under the loose bark of fallen trees. In days gone by, when the only source of fuel was wood, it was more than likely that many a salamander was borne away into the wood piles amassed for winter burning; as they are actually allergic to dry heat, the log in which one was concealed would barely have settled on the hot ashes before the unfortunate creature made a belated bid for safety—and here we have a salamander actually appearing out of the fire, foundation enough for any legend in an age when scientific reasearch was as yet unheard of.

Snakes have crawled through the pages of our history books

When we turn our attention to the reptiles, snakes are probably the most fruitful subject in these particular fields. Even today, when universal education is almost world-wide, the ignorance, superstition and fear surrounding these animals are quite incredible, and from this we can learn a good deal about the strange mentality of man. I suppose, as far as western civilization is concerned, it all started with Adam and Eve, for I believe that much of the horror evinced towards the 'serpent' even today stems from its ineradicable image as 'chucker out' from the Garden of Eden. Of some 2700 species of snake known to science, no terrestrial species is slimy and only about 150 are sufficiently venomous to harm a healthy human adult; they do not 'sting', and the tongue which flicks in an out is merely a sensory organ; they are for the most part placid, peace-loving creatures, asking only to be left alone to go their ways unhurried and undisturbed, yet, even in this so-called enlightened age, I estimate that 95 per cent of people regard snakes with fear and horror; Eve has a lot to answer for. Throughout the ages, however, they have always exerted a strange fascination over their human relatives. From Medusa with her serpentine locks right down to the insignia of the British Medical Association, snakes have crawled through the pages of our history books, as omnipresent as war or famine, or scientific progress. Choose almost any subject, age or country you like, and you will be almost certain to find a snake of some kind or other twined inescapably around it; the species depicted by the BMA, incidentally, is the Aesculapian Snake, still to be found in isolated colonies throughout Southern Europe on the sites of former temples to Aesculapius, the Greek god of medicine.

We could do worse than hang a lesson in social science on the homely bantam. The Red Jungle Fowl of India was the wild ancestor of all domestic poultry; picture a smart little Rhode Island bantam cockerel strutting proudly through the jungles with his train of dowdy little wives. And how ingenious is man; he captured the wild birds, tamed them, and induced them to lay eggs for him—so much easier than hunting the forests for these delicacies, and he could soon obtain them out of season too. The birds' carcases came in handy as an easy meat supply as well; he had feathers for decoration or, later, to stuff his pillows; and he could breed as many as he needed at will. He could breed them—that gave him ideas, and he started to select his breeding stock very carefully, choosing the parents of new generations of chickens for their size and colour, fullness of flesh and free flow of eggs. And, in time, from this single species, he has managed to produce both the decorative Frizzles and Silkies and the heavy Marans and Buff Orpingtons, the pure white Wyandotte and the Black Leghorn, the grotesque Bareneck and the Yokohama cock with a tail up to 8 metres long! A history lesson indeed—and all included in one commonplace, backyard domestic bird!

For our sample mammal let us look at the possibilities of the cavy. Probably descended from the now extinct Restless Cavy

(*Cavia pamparum*) of Brazil, this is obviously an animal of the pampas. With its short legs, almost non-existent claws and total lack of tail, it is quite obviously not adapted to burrowing, scrambling up rocks, leaping from ledge to ledge on a mountain side, or travelling or changing direction at great speed. The only condition under which an animal which is obviously prey rather than predator could survive with all these short-comings is the thick grassland of the savannah. Only here could it find immediate shelter by scurrying a few short paces in any direction; only here would its corpulent person prove no barrier to progress from one feeding area to another. Furthermore the precocious young, born exactly like miniature adults and able to run about and nibble solid food within hours of birth, are typical of the herbivorous mammals to be found in such a habitat. Here again our zoological detective-work can readily conjure up a realistic and lively picture of the South American scene. But what of their history? The American continent was discovered, as we all know, by Christopher Columbus at the end of the fifteenth century, and by the time Sir Walter Raleigh returned triumphant across the Atlantic with his deadly cargo of potatoes and tobacco the Spanish colonies were already becoming well established there. It was not long after this that the first cavies were brought across to Europe—not, surprisingly enough, as pets, but as an item to add to the menu!

Which animals to select as my examples?

As has already been stressed more than once, it is impossible within the confines of this comparatively short book to do more than offer a few suggestions as to the possible uses to which animals can be put in the educational field, and in this chapter I was quite 'spoiled for choice' as to which specimens to select for my examples. The domestication of the canary from the wild Serin Finch or the budgerigar from the Undulated Grass Parakeet would have offered me equal scope, as would the colonizing feats of the brown rat, the marathon journeyings of the Common Eel, or the accurate reflection of their type of food plant to be found in the superb camouflage of the various species of stick insect. But it does not stop even there. By deliberately restricting ourselves to only a few of the animals in primary schools, we have *a priori* reduced the animal kingdom from a million species to no more than a few dozen. Remember the other 999,950 odd, who may be brought into your classroom temporarily and spasmodically to provide yet further 'pegs' upon which to hang any line of teaching you may wish to follow.

At Paxton Junior School in South London an attic has been converted into an animal room

ARTS, CRAFTS, MUSIC AND CREATIVE WRITING

Every time I visit a school to talk about and display the animals I take with me, I am aware of the wealth of drawings and compositions which are the inevitable 'follow up' work of such a lecture, and which are, of course, exceedingly valuable, teaching as they do both observation and self expression. Although the average

A sample of work produced by 4th year pupils in a Yorkshire school

child's drawing of, say, an owl leaves much to be desired from the point of view of either art or accuracy, the very fact of having to do that drawing does make him really look at the subject and see form and colour in much greater detail than would otherwise have been the case. Similarly the effort to describe in words what he has seen evokes a beneficial effort both in memory and the use of language. Even in this comparatively restricted field much can be done. By using the classroom pets as models children can be taught not only to observe, and to some extent reproduce, form and colour, but also the use of shading and highlights (try a close-up of one of the animal's eyes, for instance), perspective (sketching the same subject from different angles) and, above all movement. It will tax the ingenuity of even the most artistic eleven year old to think up ways of depicting a gerbil 'drumming' with his hind feet as a 'still' picture! From this, of course, we can progress to models in clay or plasticine, caricatures devised from pipecleaners and papier mâché, or even more life-like figures carefully clad in hair and feathers saved from the animals' own castings. Let the children make collages from such materials too; the sloughings of stick insects, snakes and lizards, the discarded 'homes' of caddis fly larvae, feathers and hair from the birds and mammals, fragments of egg shell and, of course, the prepared bones, teeth, skins or tortoiseshell plates of deceased animals all provide excellent subject matter for this type of work. Animal portraits can be devised in appliqué work of cloth, felt, or even sticky paper, glued or stitched into position to make almost anything from a straightforward picture to table mats and Christmas cards. Or alternatively quite madly original animal creations can be constructed from paper, face flannels, felt, or more natural materials such as shells and fir-cones.

But let us not ignore the animals' own craft work. How skilfully that caddis fly larva constructs his own little shell from whatever materials may be available, gumming scraps of sand and leaf and pebble together with his home-made cement to fashion an amazingly strong and beautifully camouflaged little fortress within which his soft body is safe from most of the dangers likely to beset him. Admire the Siamese Fighter's bubble nest, and the more intricate weaving of the canary. Look closely at the finely shredded paper with which the mother hamster makes an enclosed ball-like bed for her babies and how the rabbit lines her mounded nursery with soft fur from her own breast. If we look outside the classroom to the world of animals beyond, we shall find that most of our crafts were in common usage there long before we thought we

How skilfully the Caddis Fly Larva constructs his own little home from whatever materials may be available

'invented' them. The weaver and tailor birds wove and stitched long before the first loom or needle was thought of; the beavers felled trees and built their dams; bees and flamingoes fabricated their nests in wax and clay. You name it, nature has it; even the spider's web is a close approximation to the glass fibre which is our pride and joy.

Interesting creations can be produced using simple materials such as flannel, felt, wool, feathers, sea-shells, horse-hair and the odd touch of paint

It is easy to see how the visual arts and crafts can be latched on to animal life; but how about creative writing and music? Of these two the former is the more simple—a straightforward description of any of their pets will certainly tax the pupils' ingenuity and make them rack their brains to put into words the effects it evokes. Let them try their hands at comparison—metaphor and simile—and you will come up with some truly shattering results. I have read essays by ten-year-olds mentioning the 'bent leaf turn' of a soaring kestrel, whose 'smooth strange body streamed thro' the air', the 'lickrish tail' of a vole with 'feet like wire twisted' and a 'fat, gray, brown, smooth body', in his 'running tunnel'—and these turns of phrase were not the work of a single budding prodigy—each has been chosen from different compositions on the same subject, the only common denominators being a teacher with imagination and a subject which stimulated the authors' interest and enthusiasm. With younger children you might try getting them to write down lists of words which a certain animal brings to their minds, or ask them to find comparisons for various aspects of it—hair like . . . ?, eyes like . . . ?, beak like . . . ? and so on. Colour, texture, form and movement can all be described verbally, providing an exercise at once educational and enjoyable.

As for music—the animal kingdom is full of it—music which many composers have heard and emulated. Let the children listen to Mendelssohn's *Bees' Wedding* as an example and then let

As for music, the animal kingdom is full of it

them loose with their recorders to copy the sounds made by their pets. Listen to the impeccable rhythm of your gerbils 'drumming' or the hamsters gnawing—and ask yourself and your pupils what pop group boasts a percussionist in the same class. The canary's song is legendary, but how about the sonorous booming of the tree frog? In point of fact nearly all our musical instruments have their counterparts in nature. In an imaginary animal orchestra many bird songs can represent flutes and other wind instruments; the drumming snipe, thumping rabbit or booming emu lend a variety of drums; crickets and grasshoppers strigilate on their violins, and even the humble triangle is aped from the throat of the bellbird! Elephants and mice, lions and lambs all have their individual calls which can be likened, with a little imagination, to almost all the musical sounds devised by man, and it is surely by having their imaginations roused and senses alerted by such analogies that children can be led truly to appreciate music, not merely as an aesthetic value but also as an emotionally evocative experience.

PHYSICAL EDUCATION

I often think that other animals display more straight sense through their innate knowledge (instinct) than we do through what we choose to believe are the highest of intellectual thought processes, and this is particularly true when it comes to understanding the workings of our own bodies. I am constantly appalled by the ignorance of the man in the street concerning his most elementary bodily functions, and it is small wonder that so many petty ailments are aggravated out of all proportion when that ignorance is boosted by a plethora of superstitions which one would think should have been quashed long ago in our so-called scientific era. One of my first monkeys, for instance, was given to me by a young couple who sincerely believed that the baby they were expecting would be born with simian features if they did not get rid of their pet. What they thought *my* children would look like I shudder to think!

It is generally accepted, if somewhat vaguely, that 'exercise is good for you', but I feel it important that PE teachers should make it clear to their pupils just *why* certain forms of exercise have a certain value. The fundamental fact is that, whereas use gradually destroys inanimate matter (a car's engine will 'wear out' more quickly the more it is used), it actually strengthens living tissue. We all know how feeble our legs feel after a week in bed —and the same laws apply to all the other parts of our body.

Watch a mouse on his wheel or a ferret trotting rapidly up and down his cage. No, these animals are not 'poor caged creatures' going crazy in their imprisonment; they are simply doing their 'daily dozen' which they know instinctively to be essential to their physical well being.

Notice what parts of their bodies different animals exercise. In all species the limbs, of course, must be kept in tip-top condition for flight, chase, or merely perambulation—hence the mouse on his wheel and the ferret on the trot. But look at the rabbit's nose, the chinchilla's whiskers. Always 'on the go' aren't they? Should the muscles controlling them grow sluggish or atrophy the animal's very life might become endangered because they could no longer operate swiftly enough to waft the scent of danger to his waiting nostrils. A dog gnaws a dry old bone both to clean his teeth and exercise his jaws. A cat 'kneads' a cushion and tears at the furniture not only to sharpen her claws, but also to keep the tendons and sinews of her legs strong and flexible. When our pets scratch vigorously our minds tend at once to thoughts of fleas, but in fact a good deal of such scratching is no more than massage to keep the skin fresh and the muscles supple. How often, except in the case of a few highly domesticated animals who have, alas, been infected by their masters, does one see another animal, even in middle or old age, with sloppy, flaccid stomach muscles, a sagging jaw-line, a drooping double chin? Our ancient nanny goat wears no corsets, yet her figure bears comparison with that of her grand-daughter. Basically it is only because we jump into our cars

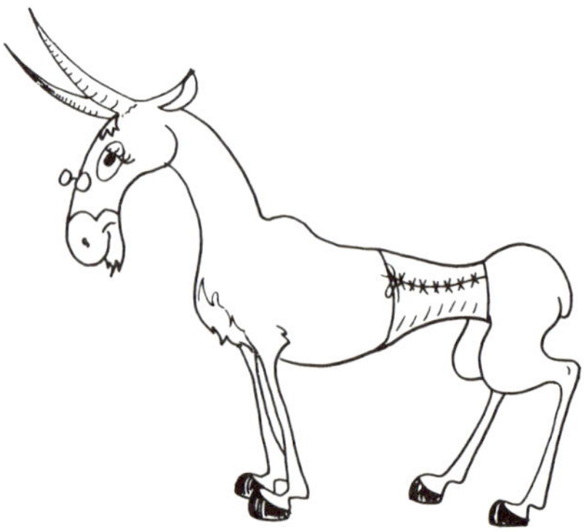

Our ancient nanny goat wears no corsets

instead of walking to work, and buy our apples from the supermarket instead of climbing the tree for them that we need to do exercises at all. If we walked and ran, stretched and twisted, climbed and crawled, and ate tough, raw foods, exercise would be no problem at all but the natural concomitant of a way of life.

But we don't, nor are we likely to; so back to the gym we must go to limber up and find new tone for the muscles and sinews which would otherwise soon atrophy.

Fortunately most children enjoy exercise; left to their own devices they are forever running and jumping, turning somersaults and standing on their heads, and such forms of disciplined exercise as games and vaulting find instant approval. More difficult to encourage with anything like enthusiasm are the more static movements such as press-ups or 'bicycles', but even these can become both sensible *and* good fun if their value is made clear while at the same time they are related to the animal kingdom. Let them pretend to be crocodiles, whose every movement on dry land is necessarily performed as a series of press-ups owing to the angle at which their limbs are set to their bodies—more like our own arms than the legs of any other quadrupeds. Let them look closely

In P.E. press-ups are hard work, but our Monitor Lizard shows the children how easy they really are

at the forelimbs of their salamander or lizard, then imagine how it must feel to propel a vast, much heavier, body by such an ungainly mode of progression. They can develop their leg muscles by pretending to be toads or rabbits, emulating whichever appropriate pet they may have available for observation, by leaping upwards and forwards from the crouching position.

Any snake makes an excellent object lesson in muscular control. Hold your specimen gently upside down by the root of the tail and watch how he pulls himself up on his own body—the Common Boa is probably the best example, as he tends to move in a slow and measured manner, raising his own weight with a minimum of fuss and apparent effort. Then see if the children can do the same thing hanging by their knees from a horizontal bar.

They can try walking on all fours like a cavy does (diagonal arm and leg moving together) and like a salamander does (arm and leg on the same side of the body moving together—much harder, and bringing far more muscles into play).

I suppose speed and bodily efficiency are among the most obvious lessons we can learn from other animals. Let the children listen to a hamster gnawing (like some madman frenziedly sharpening a pencil!) or a gerbil 'drumming' with his hind feet—and then see how quickly they can move their own legs and jaws. Watch a rabbit twitching his ears—can they do the same? If they can, or can learn to, the effort will benefit their facial muscles enormously, just as flexing their toes independently will benefit those in their feet and ankles—watch the same rabbit's distended toes as he industriously washes a hind leg. Let the pupils lie flat on the floor and strive for overall suppleness by twisting and wriggling the whole torso emulating a snake or a fish, or trying to roll up into a tight ball like a sleeping hamster.

Really the possibilities are almost endless, but the one basic fact remains unalterable; when it comes to efficiency of muscles and hence of movement, other animals are the best teachers; so, whatever pet or pets you have available, take them into the gymnasium with you as living mentors and let the children, by emulating their every activity, gradually begin to acquire something approaching the same degree of co-ordination, symmetry and muscular efficiency.

ENGLISH AND RELIGIOUS EDUCATION

If you walk round the children's section of any bookshop you will be surprised at the high percentage of volumes which deal with animals. There are factual books and fictitious books; books of animal photographs, books of animal cartoons; books of animals masquerading as human beings. From Rudyard Kipling and Beatrix Potter down to Walt Disney, animals have always appealed to the child mind at all levels; even the most old-fashioned 'first reading books' apparently cannot do without them—'The cat sat on the mat', 'Mother, look at Rover's hat', and so on. Animal alphabets are common in pre-school books; tales of horses especially are churned out by the dozen to delight the female teenager. In fact, if one so desired, it would be not only possible but quite ridiculously easy to supply a progressive booklist from the nursery class to sixth form secondary school with animals alone as the central theme. I am not suggesting that this should in fact be done; as was made clear earlier, a surfeit of *anything* is undesirable. But two facts remain basically true: a young child will learn to read much more quickly if he finds the subject interesting, and most young children are quite fascinated by other animals.

Literature, like all the arts, is an interpretation of life, and should be taught as such. Parsing Shakespeare's grammatical constructions, listing his similes and metaphors, analysing his variations upon the iambic pentametre are all very well, but they tend, I fear, to make the student regard his works as examination exercises rather than as the thrilling, living, wonderfully evocative and imaginative creations that in fact they are. Literature holds the mirror up to life; let us then teach the life of literature rather than its dissected physiology. For the very youngest readers we can easily have Peter Rabbit, Hunca Munca and her friends, Jeremy Fisher, Samuel Whiskers and Anna Maria actually in the classroom; even Mrs Tiggywinkle and Squirrel Nutkin may occasionally come our way, while Sally Hennypenny and Jemima Puddleduck may well be living out in the school grounds. If your pet mice are sufficiently tame, you can let them romp through the nursery class's dolls' house, or at feeding time explain that lettuce really does have a 'soporific effect' on young rabbits (the laudanum content may even prove fatal if they have too much). Many of the *Tales of Beatrix Potter* can literally be brought to life if you choose the right classroom pets.

Winnie the Pooh, The Wind in the Willows, and *The Jungle Book* obviously present more problems; even I would not suggest that you fill your classroom with bears and tigers or badgers and moles! But, by drawing analogies with the animals you *have* got you can help to bring them to life in a way that would be impossible without your pets' assistance:

Ye are better than all the ballads
That ever were sung or said,
For ye are living poems
And all the rest are dead.

Read the description in the *Just So Stories* of the cat walking on his wild lone, or that in the *Jungle Book* of Kala Nag lifting up one forefoot and then the other, and bringing them down on the ground 'one-two, one-two, as steadily as trip-hammers', and see if the children can think up equally vivid descriptions for the movements of their hamsters or goldfish. We cannot have Rikki-Tikki-Tavi in the classroom, but the ferret which we can have will provide an excellent substitute so far as movement and, to a large extent, behaviour are concerned. Once children realize that all the greatest literature is about *real* people, *real* events, even *real* animals, albeit translated into the realms of fantasy, they will begin to have their first inklings of what it is all about.

Rather along the same lines, it has often occurred to me that the Bible might be more widely read if it were presented in school as a literary work, which of course it is, rather than as a highly moral treatise on religious teaching. Basically, few of us care much for moralization; to the average person nothing is much more irritating than having his own standards impugned and a different system of morals thrust down his throat. If, then, we can bring the Bible to life for our children in the same way that we can *The Wind in the Willows* and Mickey Mouse, we shall have gone a long way towards making them not only appreciate it as a work of art, but also pay serious attention to its message.

Here again, our animals can help us, for throughout the Bible, especially perhaps throughout the New Testament, there runs a

21 *If, then, we can bring the Bible to life. . . .*

steady theme of protectiveness and kindliness towards the rest of the animal kingdom. 'Shall two sparrows, sold for a farthing, fall to the ground unnoticed by your heavenly Father?'. 'Ask now the beasts and they shall teach thee'—yes, the whole theme of this book might be said to stem from the Bible, and indeed a wealth of animal life wanders steadily through its pages. One of the drawbacks here, of course, lies in the variety of interpretation, for different translators have given quite different identities to many of the creatures mentioned, the Authorized and Revised Versions seldom coinciding. Therefore what appear in the Authorized Version as tortoises, ferrets, chameleons, snails, moles and spiders, in the Revised Version are referred to as great lizards, geckos, crocodiles, sand lizards, chameleons and lizards! One would deduce that the translators of the former were more imaginative than those of the latter! Animals make their appearance in the Bible under many guises; they are often upheld as models for human behaviour ('Go to the ant, thou sluggard'), used to signify the wrath of God (the plague of locusts), or put to work as the instruments of punishment (the early Christians being thrown to the lions, or 'I will chastise you with scorpions'). Sheep, especially rams, were in great demand for sacrificial purposes and, of course, together with goats, played a vital part in the economy of a nomadic race; Elijah was fed by the ravens and a dove brought tidings of dry land to Noah; the ox and the ass worshipped the infant Jesus, who was born among them in a stable. From the serpent in the Garden of Eden down to the cock who crowed Peter's betrayal, animals figure throughout Old and New Testament alike, used directly, used in metaphor and simile, to illustrate and enliven the religious message. Indeed, the many writers from Moses to Saint John himself who put pen to paper to tell the story of a faith which, over the centuries, has swept the world, apparently realized the value of animals as a teaching medium, and used them accordingly. It might also be worth drawing analogies with other religions, for even the most primitive tend to incorporate animals and other natural phenomena into their rites and beliefs—from the man-beast werewolf and Leda's swan to the Buddhists, certain of whom even today sweep the ground before them lest their step accidentally bring death to any of God's smaller creatures. 'If he shall ask an egg, will he offer him a scorpion?'; 'Behold, you are beautiful my love, behold you are beautiful! Your eyes are like doves behind your veil. Your hair is like a flock of goats moving down the slopes of Gilead'; the swallows have a place to nest, 'even thine altars, O Lord of hosts'. Yes, animals have a firm footing in the Bible, both in fact and symbolism, and through them we may well be able to bring the old story to new life for our children.

Animals abound in literature from the Bible to Enid Blyton; but I wonder if you have ever thought how insidiously they have crept into our everyday language too? Most children have at one time or another enjoyed working out the quizzes in their comics which demand the completion of a well-known simile—eyes like a . . . ?, blind as a . . . ?, sly as a fox, work like a beaver. Your class can learn a great deal about the English language by listing as many of these as they can or completing those you start for them. Animal endearments (pet, chick) or insults (dirty dog, pig, cow, swine) are familiar to us all, and their zoological origins are quite obvious, but we use many more subtle allusions every day without, I suspect, ever thinking of what they really mean. When we 'ferret something out' or 'worm our way into this or that', when we 'monkey about' or 'rat' on somebody or refer to an acquaintance as 'mousey' or 'catty', how many of us really think about what we are saying and allow the mental picture these phrases conjure up actually to appear before us? Even more unobtrusively animal in origin are many words, verbs especially, which have become such a commonplace in our conversation that we no longer think of them as metaphorical at all—to howl, to gallop, to trot, to hatch (a plot) or to crow are all in daily use as words in their own right without the faintest flavour of their first use as imagery to enliven the bare bones of meaning. How many of us, when we say 'Don't crow over me', actually envisage a cockerel triumphant on his dunghill? Yet what an apt allusion this is when we *do* think about it! In order that they may be able to express themselves intelligibly and correctly, it is essential that children be taught the rudiments of grammar; they must indeed be able to parse, to distinguish a noun from a verb and an adverb from an adjective, to seek out the subject of a sentence—and make sure that what they write *is*, in fact, a sentence at all! They must be able to spell and punctuate and know when capital letters are required—but, above all, if they are to be able to put into words the experiences they want to pass on to others, they must learn their language as a *living* thing, full of vitality and innuendo and above all endless possibility. Let them see words as their tools, not their masters—tools which have so many facets and uses and intricate shades of meaning that nothing on this earth is beyond their powers of expression, and the first step towards this depth of understanding is for them to find out just exactly what each word means or can be construed as meaning. When we 'flounder about' what are we doing? We visualize ourselves slipping and struggling on a wet and muddy surface, flapping our arms about as we strive to retain a precarious balance, making heavy weather of it (a nautical allusion here!) as our sliding feet are sucked yet deeper into the mire —in fact, behaving very much as would the flat fish from which this particular verb is taken when he is hauled unceremoniously from his native element and dumped, gasping and struggling for breath, upon an unyielding deck. And all this suggested in one short word.

Obviously you are unlikely to be keeping flounders—or, for that matter, monkeys, worms, foxes, or beavers—in the classroom, but I know of no words in common parlance derived from budgerigars, gerbils or hamsters. You can certainly study your mice, ferrets, cockerels, fish and leeches in relation to their metaphorical usage, but this leaves you somewhat limited. How about, then, letting the children watch any one of their pets really closely, preferably intermittently over a period of days or even weeks so that every aspect of its behaviour becomes familiar to them—and then getting them to write about it? Let them, in fact, produce their own allusions and metaphors by relating their acquired knowledge to new experience—they might even come up with a brand new word, or use of a word, which will eventually find its way into the Oxford Dictionary!

AGRICULTURE AND ECONOMICS

The origin of the domestication of one species of animal by another is sunk in the mists of time. Long before the first man brought a wolf into his cave and called it a dog, ants were keeping herds of greenflies and 'milking' them much as we do our cows; as I have said before, nature got there first in the case of most of our bright ideas. Certainly, though, man has domesticated more different types of animal than has any other species, and, indeed, much of our economy is dependent upon them. In rural communities a reasonable proportion of the pupils at any school will eventually find their way on to the land, so it is hardly surprising that here at least agriculture in some shape or other should feature upon the curriculum.

Although we are no longer dependent upon the horse for our transport or the dog for our hunting, we have not yet, thank goodness, found an artificial substitute for milk or eggs or a leg of mutton, so farming is likely to remain a major part of our lives for a good many years to come. Consequently it seems to me that the sooner budding agriculturalists are introduced to the practical side of animal husbandry the better. Many secondary schools in country districts have, in effect, their own small farms and, while cows and sheep would be likely to pose problems for the primary school, there seems to be no valid reason why goats and poultry should not be kept on a realistic basis. Leaving aside arable farming—for, unless the school has a great deal of land at its disposal, the production of any sort of crop even for the feeding of its own livestock would be out of the question—we are left with two main aspects of agriculture: husbandry (care of livestock) and economics. Fortunately the two can and should go hand in hand.

The recent upsurge in factory farming is, I am quite sure, wrong, not only morally but economically in the broadest sense of the word. So far I have never met any reasonably decent person who does not consider such treatment of any living thing distasteful to say the least, and I have also met many farmers who, after a good deal of experimenting, have been forced to the conclusion that it is, in fact, cheaper to produce meat by more natural methods. On a wider view, too, the food value of intensively produced commodities, whether flesh or vegetable, is inferior to that of the same commodities produced in the time they were designed to take. As a scientist once remarked to me, a carrot simply has not time to absorb sufficient nutriment from the soil if it is grown, with the aid of artificial fertilizers, at the rate of two crops per year instead of one. Certainly, in common with anybody who has cared for animals for a number of years, I can confirm that, whereas two carrots would have provided an adequate daily ration for a rabbit ten years ago, three or four are required now if the animal is not to perish from malnutrition. Presumably the same applies to meat, eggs, milk, etc.

Let us restrict ourselves, then, to the type of farming which has stood the test of time if we incorporate it into our school syllabus. Goats, bantams, ducks and geese have already received a mention as animals suitable to be kept in schools *outside* the classroom; let us see what educational use can be made of them. Apart from their obvious value as animals to be looked after, yet more examples of creatures other than themselves which the children can be responsible for, they can form the basis for a course in agriculture which might well be prolonged throughout their whole school career. Even an eight-year-old can learn to milk a goat or to collect the eggs from the ducks each morning and from the bantams in the afternoon. They can learn to filter and cool the milk to prevent its going sour too quickly, and, if the County Authority is

Even an eight year old can learn to milk a goat or collect eggs from the ducks

willing to provide the necessary equipment, separate the cream and go on to make their own butter and cheese. The eggs can be washed and stacked in egg-trays or preserved in waterglass for winter use (not the duck eggs, though, as the shells are hydroscopic and thus preclude long storage), or left uncleaned for sitting under broody bantams to produce more stock. Goats, of course, must breed in order to produce milk, so the mating of the nanny and birth of the kids may well be incorporated into our lessons on the facts of life.

Nowadays, when money looms so large in the lives of most of us, even quite young children seem to find additional satisfaction from anything that 'pays', and in this field they can start their first lessons in profit and loss. Obviously one goat and half a dozen bantams are not going to make anybody's fortune, but even on the smallest scale it should be possible for them at least to pay for themselves. If you encourage the children to sell their produce —eggs, milk, butter, cheese and, eventually, surplus stock, and at the same time insist that they buy their animal foodstuffs from the money thus acquired, it will be surprisingly easy to introduce them painlessly into the realms of book-keeping—especially if they are allowed to put their small profits towards some other favoured project. I know a few schools who run quite a thriving little business along these lines, some of the secondaries even taking their wares to market in the age-old approved style; in primary schools, however, I feel that the customers had perhaps better be restricted to parents and friends—although an eye should be kept on the possibility of inflationary prices being charged to doting mothers!

In all walks of life it is an inescapable fact that you will get out of anything only what you put into it, and the same rule applies to animal husbandry. It is no good hoping to bump up profits by skimping on feed; a goat's milk yield will always be in direct proportion to the concentrated food it takes in, and similarly no duck or chicken will produce its maximum number of eggs on scanty rations. You can prove this graphically without harm to the animals by, for example, restricting the goat's diet to her browsing, hay and the merest handful of dairy nuts; her own physical well being will not suffer as she will be eating plenty for her bodily needs, but her milk yield will go down with a bang—and another useful lesson will have been vividly taught.

It is no good hoping to bump up profits by skimping on feed

If you wish to develop this theme yet further, you could, of course, start 'farming' other animals, such as rabbits. Many of the children, and, I suspect, not a few of the teachers, will not be too happy about breeding animals for meat or pelt production—and I must admit I should not much care for the idea myself. But certain breeds of rabbit, notably the Angora, can be kept for their wool, which has quite a high market value if properly produced, and all, of course, can be bred to supply the pet trade, as can mice, hamsters, gerbils, budgerigars, canaries, pigeons, tropical fish and others. I feel, though, that one should guard against getting quite carried away by the 'farming' aspect and becoming *too* mercenary; while economics must necessarily, alas, play quite a large part in the lives of us all, it is as well to remember that there are other things in life of equal importance, and an appreciation of animals for their own sake is not least among these.

NATURE STUDY

There are two main facets of nature study at primary school level which may be investigated separately or concurrently—field-work and classroom study. In country schools the 'nature walk' has always been an intrinsic part of the syllabus, with the gum-booted and rain-coated children happily indulging in pond dipping, collecting flowers and fruits and grasses from the hedgerows, or simply being taught to look at the world around them. Provided that they really *do* look, and duly classify their trophies, this is of obvious value, and can easily be developed into an ecological study to give the pupils at least some idea of the intricate and subtly balanced relationship between animals, plants and minerals, each constituting the other's environment. In fact it is *just* possible (but only just, alas!) that such studies might help to produce a generation less inclined to fly blindly into the face of nature than we are at present, and thus avert that complete disintegration of this planet which twentieth-century man seems so grimly bent on achieving.

On the zoological side, pond dipping is the most obvious extramural activity, as it has the advantage of continuity—since live specimens can be brought back into the classroom with a fair chance of continuing their life cycles under observation and even under experimental conditions. The natural law of camouflage, for instance, can be seen in operation if we periodically change the habitat of our caddis fly larvae and watch them vacate their twig-like residences for fresh quarters constructed from and thus perfectly matching whatever floor covering we care to introduce into the aquarium—sand, gravel, dead leaves or even such artificial building materials as fragments of glass or pottery or plastic will all be adopted with equal ardour by these indefatigable seekers after anonymity. If we arrange our pond life tank carefully, with sufficient living plants to ensure a steady supply of oxygen, predators and plant-eaters, street-cleaners and filtration machanics in the right proportions, we can see in miniature the perfect balance of natural economics which is capable of ensuring the harmonious and smooth running of any natural community. In a comparatively small classroom aquarium one dytiscus beetle would be one too many, but a single small leech might just keep a fair sized colony of pond snails within bounds, while they in turn painstakingly ate the algae off the glass, a Swan Mussel ploughed slowly across the sandy floor filtering the water as he went, and the caddis fly larvae and water boatman disposed of any bodies, animal or plant, which had died from natural causes.

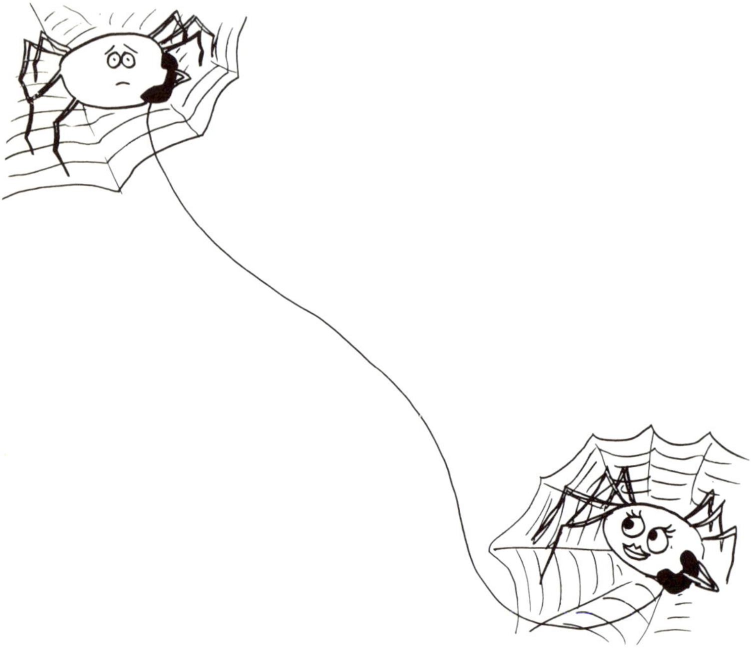

The ardent suitor takes suitable precautions by 'ringing up' his beloved

We humans are inordinately proud of our economy and especially of our 'inventions'—but virtually all of them were exploited by other members of the animal kingdom long before we even made our first tool. The Swan Mussel and many other aquatic molluscs developed the principle of perfect filtration far in advance of any sewage works, and the closely related snails with their 'endless belt' of 17,000 teeth surely forestalled the band saw just as the land varieties forestalled our track-laying vehicles as they spread their own 'roads' of slimy mucus. Squids 'invented' jet propulsion long before the Wright brothers' first attempts at aviation and, of course, the bats' sonar devices are common knowledge. Not so well known, though, are the telephones and liquidizers, computers and chronometers, used by other invertebrates. The female spider's habit of devouring her spouse is famous (or should I say infamous), but in one species at least the ardent suitor takes suitable precautions by 'ringing up' his beloved to ascertain her frame of mind before embarking upon his dangerous courtship. He strings a slender line from his web to hers and then, from the safety of his own stronghold, strums hopefully upon it and judges from the ferocity of the return message whether or not it is the right moment to press his suit! The scorpion, having immobilized his prey with a tranquillizing injection, proceeds to mince up and finally liquidize it with his chelicerae (mouth parts) before ingesting the resulting 'soup' into his digestive tract. The foraging honey bee, having received detailed instructions from her fellow workers as to her exact target, together with its direction and 'distance off', will fly unerringly as much as three kilometres to the precise source of nectar (or pollen, or propolis, or water) of which she has been informed, and later, having 'tanked up', return just as unerringly to her hive. When one remembers that the sun by which she steers moves 15° in an hour, and that she may be away from her hive for even longer than that time—one begins to wonder just why we are so proud of our time-keepers, computors and calculating machines! Even such apparently purely human inventions as night-storage heaters have their counterpart in nature—for what other reason do reptiles bask long hours in the sun than to absorb and store heat to tide their cold-blooded bodies over the chilly night to come?

Another quite different method of approach is to select a single animal, either one of your classroom pets or alternatively a wild one easily observed in the field, and explore all the other subjects on the syllabus via this solitary medium. Let us take the Mongolian Gerbil (*Meriones unguiculatus*) as a pattern. Obviously a geography lesson on the chill wastes of the Mongolian plateau can easily be drawn from this charming little rodent, and its long and chequered history is a natural concomitant. From the great Mongol ruler, Timur i Leng (Timur the lame), immortalized in Marlowe's 'Tamburlaine the Great', we can step easily into the realms of Elizabethan drama and thence, if we so wish, through the whole history of English literature, coming full circle back to our original theme with Coleridge's 'Kubla Khan'. Or we can plunge back into pre-history if we turn to the discovery of Dinosaur eggs near Shabarakh Ussu in the early 1920s. On a slightly

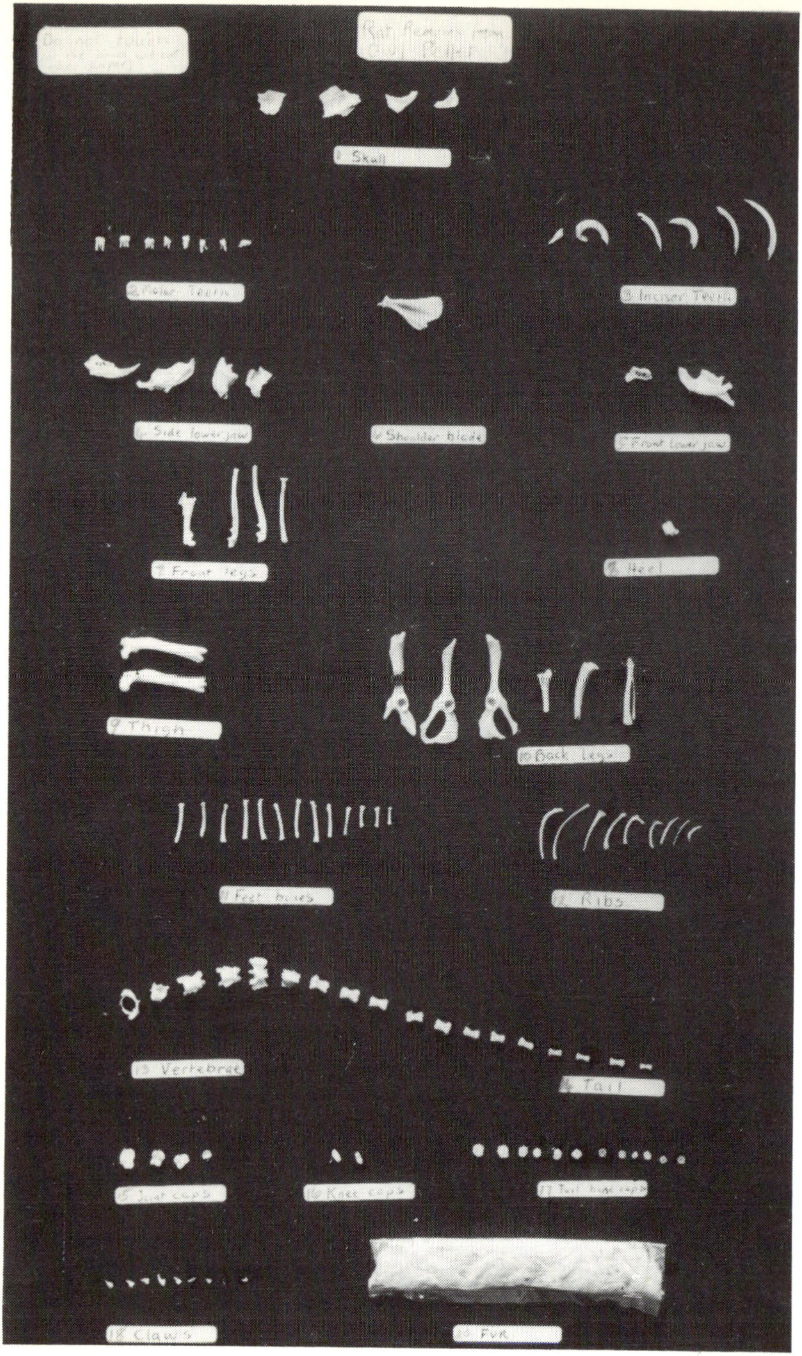

Keep a lookout for owl pellets when on walks in the countryside. These can be dissected and their contents identified and displayed on black card.

different tack we can note that the closely related Libyan Jird (*Meriones libycus*) is mentioned in the Bible as the 'desert rat'; 'desert rats'?—were not Montgomery's forces in the Second World War thus designated? Of course they were—and here we are, back in the realms of history once more, only this time concerned with our own twentieth century! Even a touch of medical knowledge can be introduced from our hard-working classroom pet, for are not certain degrees of mental disorder currently known as 'mongolism', owing to the slanting eyes and high cheekbones which typify the condition, physical characteristics gradually developed among the Mongolian tribes to protect theur eyes from the glaring skies and stinging sand-filled winds of their native deserts. I know of one school where a whole term's work was based on a single country—geography, history, mathematics, nature study and all the other subjects being geared to the same topic—and it seems to me that the adoption of a classroom pet as the basis for such a project, the living symbol of all the work in progress, could do nothing but add to the value and reality of the enterprise.

If we select a wild animal as our theme, care must be taken to choose one frequently seen and fairly easy to study. We can follow the rabbit, for instance, through literature (Beatrix Potter's immortal tales), across to Australia (history, geography, sheep-farming) and let it teach us among other things the danger which lies in uninformed meddling with nature. We can compare other introduced species from the Fallow Deer brought to our own country by the Romans to the dingo in Australia and the Grey Mongoose in Jamaica.

All knowledge is inter-related somewhere along the line, and if you first arouse the children's interest by introducing them to an animal and then let your imagination run riot, there is no telling what paths of learning you may find yourself pursuing. Only one thing is certain—information acquired by this method is acquired both painlessly and permanently, for the absorbing fascination of tracing the myriad ramifications of *any* subject tends to imprint them upon the memory for all time.

List of suppliers

ZOOLOGICAL SUPPLIERS

J.G. Animals, Streatham Vale, London SW16
(mainly reptiles and amphibians but also supply unusual small mammals, animal cages and foodstuff)

Regent Pet Stores (Palmers), Parkway, London NW1
(most classes of animals, cages and foodstuff)

Great Goodwin Farm, Merrow, Guildford, Surrey
(birds, rabbits, poultry and goats)

Parslows Hamster Farm, Common Side, Great Bookham, Surrey
(hamsters and other small mammals)

Slater Livestock, 18 Pier Road, Erith, Kent
(tropical fish, birds, reptiles, amphibians)

Drs de Rover, c/o Alan Cook, 'Anton', Beechwood Drive, Congleton, Cheshire
(reptiles and amphibians, occasional invertebrates)

Leslie Dyer Ltd, Brook Lane, Stretham, Ely, Cambridgeshire
(rabbits, other small mammals, poultry, pigeons)

Walter R. Smith Ltd, 39 Tib Street, Manchester 4
(cold water and tropical fish, amphibians, reptiles)

Worldwide Butterflies, Over Compton, Sherborne, Dorset
(mainly butterflies, but occasionally stick insects)

BIOLOGICAL SUPPLIERS

Freshwater Biological Association, Ferry House, Far Sawry, Ambleside, Westmorland
(freshwater invertebrates, fish, amphibians)

Gerrard & Haig Ltd, Beambrook, Newdigate, Sussex
(invertebrates, few fish, reptiles and amphibians, biological equipment)

Griffin Biological Laboratories, 113 Lavender Hill, Tonbridge, Kent
(reptiles, amphibians, small mammals, preserved zoological material, charts, unmounted collections, museum jar mounts, film strips and projection slides, skeletal material)

Readers please note that this is not a complete list, but merely a representative one; the authors cannot hold themselves responsible for any disappointments caused through purchases from the dealers listed, nor is any criticism of those omitted implied.

Bibliography

The following list of books deals mainly with the CARE of many of the creatures we have mentioned in the preceding pages of this book. Some volumes are older than others (though still in print) and preference in selection has not been given to any particular age group.

Frogs and Toads (Macdonald, 1971).

Spiders (Macdonald, 1971).

CARR, ARCHIE: *The Reptiles* (Seymour Press, 1968).

BARTH, G. M.: *Pond and Stream Life of Europe* (Blandford, 1963).

DENHAM, STEPHANIE and KEN: *Pets for Children* (Hamlyn, 1969).

DUTTA, REGINALD: *Tropical Fish* (Octopus Books, 1972).

FARB, PETER: *The Insect World* (Constable, 1961).

GRAF, DR JAKOB: *Animal Life of Europe* (Warne, 1968).

HERTER, KONRAD: *Hedgehogs* (Phoenix, 1966).

HERVEY, G. F. and HEMS, J.: *Freshwater Tropical Aquarium Fishes* (Spring Books, 1964).

JANUS, H.: *Molluscs* (Burke, 1965).

JENNINGS, TERRY J.: *Animals In The Home And In The Classroom.* (Pergamon Press, 1971).

KNIGHT, MAXWELL: *Animals and Ourselves* (Hodder & Stoughton, 1962).

——: *Talking Birds* (G. Bell, 1961).

——: *Tortoises and How to Keep Them* (Brockhampton Press, 1964).

KNIGHT, MAXWELL and GREEN, WILL: *Keeping Pets* (Brockhampton Press, 1971).

LE ROI, DAVID: *Hamsters and Guinea Pigs* (Kaye & Ward, 1969).

RATAJ, K. and ZUKAL, R.: *Aquarium Fishes and Plants* (Spring Books, 1971).

ROGERS, CYRIL H.: *Budgerigars* (Foyle, 1962).

ROOTS, CLIVE: *Exotic Birds for Cage and Aviary* (Cassell, 1971).

RISDON, D. H. S.: *Your Book of Pet Keeping* (Faber & Faber, 1964).

SANDARS, EDMUND: *Insect Book for Pocket* (Oxford University Press, 1946).

SIMON, SEYMOUR: *Discovering What Gerbils Do* (World's Work, 1972).

SIMSTER, W.: *Choosing Your Pet* (Nelson Guide, 1972).

TWEEDIE, MICHAEL: *Pleasure From Insects* (David & Charles, 1968).

WAINWRIGHT, NEIL: *Tropical Aquarium* (Warne, 1970).

WARNECKE, GEORGE: *Butterflies* (Burke, 1964).

Index